Driverless Cars: On a Road to Nowhere?

PERSPECTIVES

Series editor: Diane Coyle

Driverless Cars:
On a Road to Nowhere?

Christian Wolmar

LONDON PUBLISHING PARTNERSHIP

First edition published in 2018

Published by London Publishing Partnership
www.londonpublishingpartnership.co.uk

Published in association with
Enlightenment Economics
www.enlightenmenteconomics.com

ISBN: 978-1-913019-21-1 (pbk)

A catalogue record for this book is
available from the British Library

This book has been composed in Candara

Copy-edited and typeset by
T&T Productions Ltd, London
www.tandtproductions.com

Printed and bound in Great Britain
by Hobbs the Printers Ltd

Contents

Preface

It is just over two years since I wrote the first edition of this book and, if anything, my central thesis has been reinforced by events since I penned that first draft. My curiosity about autonomous vehicles – to give driverless cars their proper name – had originally been stimulated as long ago as 2013, when I read an article about them in the *Evening Standard*. My wife, who worked there at the time, had sent me a link, and after reading the piece I dashed off an email to her: 'Fascinating. This is going to happen.'

The article, by the paper's comment editor Andrew Neather, was a classic piece of futuristic optimism about how we would all soon be travelling seamlessly in driverless cars while tapping away on our smartphones or reading the *Financial Times* on 'roads packed with other self-driving pods'.

The piece was published to mark the arrival of the first 'robot-car' in Britain – a Nissan Leaf fitted with a variety of cameras and sensors – and it suggested that the driverless Leaf would soon be seen spinning round the streets of Oxford.

My hasty and unthinking response to my wife's email was very much in keeping with the zeitgeist. As soon as news started to emerge of Google's efforts to create a driverless car, a widespread assumption that we would all be using them in the relatively near future took hold. Articles began to appear almost daily in the press about the next trial of the technology, or about the investment being made by government, all in the assumed context that these things would soon become a reality. Every test, every announcement, every government initiative was hailed as a new dawn in transport technology. There was a tone of inevitability in all this coverage that soon began to permeate through to the politicians. Phrases such as 'the upcoming driverless car revolution' and 'the disappearance of the privately owned car' started to pepper their speeches. Some even assumed that the advent of driverless cars was imminent and that transport policies therefore needed to be adapted urgently. That driverless car revolution was upon us, and those who ignored it were simply doing the three monkeys trick. They were Luddites, losers. Had not the PC, the smartphone and the internet already changed our lives? 'Smart cities' were all the rage and driverless cars were one of their obvious key building blocks.

Judging by my response to my wife's email, I had clearly swallowed this line too. But then my brain started to engage. It was precisely that tone of inevitability which began to make me wonder whether this revolution in transport would really soon be upon us. Hold on a second, I thought: is this really something that is bound to happen? I began to consider the issues raised by the concept of driverless cars (or rather, as I have used throughout this book, autonomous vehicles). I began to think about the processes through which they would need to be introduced and about the implications if they were. Lots of questions came to mind. And as I sought answers, more doubts were raised. What was the technology currently capable of? Were there really driverless cars on the roads in the United States? Were they really safer? How would people react to their introduction? Why was there so much interest in the concept? What were the employment consequences?

Then there were questions about the projects and events that were being mentioned in the media. Were driverless cars really being used in Greenwich? Was it actually possible to have six lorries 'platooning' on the M6 without causing problems for other traffic? Why had Google radically reframed its test programme? What happened after the Tesla driver smashed into a lorry at 50 miles per hour when the vehicle was in 'autopilot' mode? Would it really be possible, as Google claimed, to reclaim thousands of square miles of parking lots when driverless cars were introduced? Would public transport be irredeemably wrecked by their introduction? Would the technology be affordable? You get the gist, reader, and you can no doubt pose your own questions too.

Yet the media – and the electronic media in particular – seemed to be ignoring these questions. There was simply a barrage of unquestioning techno-centric coverage that rarely – and even then barely – raised any of these issues. It's the technology, stupid – of course this will happen. There were plenty of banner headlines about 'driverless' trials that, when the background was examined, turned out to be nothing of the sort. There were no 'driverless taxis' coasting round Pittsburgh; nor were there autonomous Nissan Leafs parading on the streets of Oxford or on the highways of Mountain View. The technology companies and auto manufacturers are reluctant to describe precisely what their products are able to do, but it certainly does not match the media headlines. The world of autonomous cars is one of hype, secrecy and technological determinism that has so far not been challenged.

The *Evening Standard* was in fact rather atypical in that it did raise some crucial issues about employment and the problems with technology, and it expressed doubts about the feasibility of the concept. The Nissan Leaf has now appeared, and as an early test it went on a 230-mile drive across England from Cranfield in Bedfordshire to Sunderland, using a variety of roads including the M1 motorway and various country lanes. There were two test drivers, but it ran for 99 per cent of the time in autonomous mode. Oxbotica, the firm mentioned in the *Evening Standard* article, and the upmarket minicab firm

Addison Lee committed to putting driverless taxis on London roads by 2021 in a deal signed in 2018. There will, though, still be a 'safety driver' at the wheel, and as with so many of these trials and tests, there is much secrecy around the project and no firm news about progress. One thing that we do know is who is footing the bill for the plan: UK taxpayers are contributing £15 million for the development of these vehicles.

I started asking questions, lots of them, and an email conversation with John Adams, a geography professor and a leading theorist on risk compensation, spurred me on to raise issues that were even more fundamental. Issues such as how autonomous vehicles would cope with pedestrians on the road? Or cyclists? How would they distinguish between a traffic jam and a line of parked cars? Would they be programmed to never break the law, by exceeding speed limits for example? Or, more controversially, would they be allowed to drive illegally? Would 'bad people' or pranksters be able to stop the cars at will, or would they be programmed to run them over? Would they be programmed to drive into a wall to avoid hitting pedestrians? Would the urban realm have to be completely redesigned? In this book I have tried to provide some of the answers, or at least to frame the questions in the right way.

When I put these questions to academics working on the development of autonomous vehicles, their answers were expressed in far less confident terms than implied by the stories in the media. The advocates, when pushed, admitted that the task ahead was rather more difficult than had been assumed. There was even an admission from some that the introduction of these vehicles was taking place far more slowly than predicted by articles like that *Evening Standard* one from 2013. Two years ago I wrote, 'Indeed, at the moment there is no such thing as a driverless car', and the same is true today. There is still nothing like a driverless vehicle in the true sense of that term. My test of true driverlessness is a vehicle that would be capable of taking a passenger to their office before returning home to whisk the kids to school. We are still nowhere near that. It is decades away and it may never be possible because of some of the issues raised in this book. The obstacles – technological, social, practical, economic, regulatory, legal – are legion.

This book is not born of my love of trains and bicycles. I am admittedly an avowed and much-published fan of trains and public transport generally, but this is an objective assessment of the potential of the driverless car as a transport 'solution'. I recognize that cars have a vital role to play in our transport system, and that will remain the case for the foreseeable future. However, what is being proposed as a technological solution is, in fact, a highly political concept, and a deeply controversial one at that. As I suggest in chapter 6, there are solutions that are not dependent on the future development of technology that would better address the problems of the congestion and pollution caused by our obsession with the private car. Indeed, they could be introduced tomorrow if politicians were brave enough.

Some cities are already travelling in the right direction. London's congestion charge, introduced more than fifteen years ago, has reduced the number of cars going into the centre, and it could easily be extended. Further afield, many European cities, such as Hamburg and Helsinki, have strategies to cut car use, and Paris has announced that it will ban all non-electric cars by 2030. Equally, the reaction to the Covid-19 pandemic and the resulting lockdown has stimulated a range of radical measures to support cycling and walking. Expect many more such initiatives, which are a far more realistic response to urban transport problems than the one promoted by the advocates of driverless cars.

Chapter 1

The myth of motoring freedom

The car has dominated our lives for about a century. As late as the end of World War II, and despite the efforts of Henry Ford, even in developed nations few people could afford to buy a car. Now the 'parc' (the term for the total number of vehicles in the world) is around 1.2 billion: one vehicle for every seven human beings. By 2035 the parc is expected to grow to 2 billion: one car for every four people.

Between the wars, Ford enabled the middle classes and blue collar workers in the United States to become motorized by inventing production line techniques that cut the cost of manufacturing. Cleverly, he paid his thousands of workers well enough to enable them to buy his product, and this high-tech, high-wage model was imitated elsewhere, stimulating the mass car market. However, in Britain it was not really until after World War II that owning a car became commonplace. There were fewer than 2 million motor vehicles in the United

Kingdom in 1950, compared with 37 million today. The 1950s was a golden age when it seemed that motoring offered unparalleled freedom to travel with literally no downsides. As Steven Parissien, author of a history of the motor car, puts it: 'no one in the fifties worried about emissions, about carbon footprints or … about the ready supply of cheap oil'.[1] Cars transformed the way people lived. The very geography of towns and cities changed as suburbs, whose location had previously been limited by the need to be close to a railway station or tram stop, could spring up anywhere. Planning laws were adjusted to take into account this new-found access to mobility. The growth of low-density suburbs was a direct result of greater access to cars. Having a garage began to be seen as nearly as essential as having a kitchen or bathroom (though in practice garages more often than not ended up being used as storage spaces, with the poor car left outside in the rain).

Cars did not solely serve transport needs. They became a 'must-have' lifestyle accessory and were objects of pleasure, with different brands – *marques* – offering distinctive styles and meanings: sports cars were for fun, Rolls Royces and Cadillacs signified affluence, small cars were for convenience or for housewives, mid-range models represented suburban solidity, and so on.

Mobility came to be seen as as much of a right as housing or food. The car was the passport to freedom, and nothing could be allowed to stand in the way of that dream. Georgian terraces were felled, parks – even Hyde Park – were bisected, thousands of people were displaced, pedestrians were shunted into dingy subways so as not to disrupt traffic flow, and whole neighbourhoods were blighted by monstrous constructions such as Spaghetti Junction, the nation's grandest symbol of the deification of the car.

At first, the cost of adapting our towns and cities to the needs of the car was accepted as inevitable. Indeed, new dual carriageways, bypasses and link roads were welcomed because they were seen as a way of relieving traffic congestion. Early motorways were celebrated for ending congestion once and for all. In May 1963, after the first section of the M2 was opened, a local newspaper in Kent hailed the event in glowing terms:

> The people of the Medway towns can hardly believe their eyes. Traffic is flowing freely through Rochester and Chatham for the first time in many years. There are no long queues at the traffic lights in Strood.[2]

There are now! Neither the euphoria nor the relief from traffic would last long.

The 'golden age' was short-lived precisely because the attractions of the car were so great. The very success of the motor car was its undoing. An empty road was a pleasure, a full one a nightmare. The more cars there were, the more roads were needed; and the more roads that were built, the more cars filled them. Cars, it seems, abhor a vacuum, quickly filling up any stretch of new road, especially in urban areas. The car was a great invention until everyone else had one as well. A key research finding from a report published in 1994 by an obscure government committee (the Standing Advisory Committee for Trunk Road Assessment) was that providing extra capacity by building roads led to additional demand because it encouraged more people to jump in their cars. This phenomenon, known as 'induced traffic', means that the value of road schemes in reducing congestion is overestimated, but the recommendations in the report that the methodology for assessing road schemes should be changed as a result have been ignored.

The failure to recognize this fundamental flaw has meant that road construction has continued to be encouraged and supported financially, even though it clearly fails to deliver the hoped-for benefits. And so it goes on. In the February 2020 Budget, the chancellor, Rishi Sunak, announced a £27 billion programme of new and improved roads, and he has stuck with that policy despite the huge spending required to deal with the effects of the subsequent pandemic and despite the fact that there was strong evidence that air pollution contributed to the severity of the illness. Even Edmund King, the president of the Automobile Association, questioned whether the £27 billion of road spending was the best way to invest in transport. The phenomenon of induced traffic explains why adding extra lanes to highways so often disappoints local drivers, who find that within a short period congestion returns to previous levels.

The increase in car use developed astonishing momentum because it was effectively self-generating. As more people bought cars, the alternatives, such as public transport, became less viable, and consequently the planners' dreams became self-fulfilling nightmares. Every out-of-town supermarket development or multiplex cinema stimulated yet more growth in car use as Britain adopted the worst aspects of American planning models. Malls housing B&Q superstores, Homebase garden centres, Toys "R" Us and countless other chains spread inexorably around the country. As transport expert Lynn Sloman summarized in 2006:

> In less than forty years, the car has become so intrinsic to the way we work, shop and spend our leisure time that it is almost inconceivable that we once managed without it.[3]

Other transport modes suffered. While the cuts resulting from the Beeching report are the most infamous, the wiping out of all the nation's trolleybus schemes, all but one of its tramways (the one in Blackpool was saved), and many of its bus routes and suburban railways resulted in an ever-greater dependence on the car. Even parts of the London Underground, which saw a decline in passenger numbers, were being considered for closure. Cyclists were bullied off the roads and pedestrians were at far greater risk on busy roads where traffic was encouraged by the design of the highway to go faster. The supposed freedom offered by the car proved to be a chimera as choice became more restricted because of the decline in the use – and consequently the availability – of other transport modes. There are, as we shall see, parallels between this period of growing hegemony of the car and the future envisaged by many of the advocates of autonomous cars.

Parissien believes that 1959 represented the zenith of the car as a symbol of freedom. After that, the drawbacks began to appear, slowly at first but then so overwhelmingly that it was impossible to ignore them. A key turning point in the United Kingdom was the abandonment of the plan to build a series of 'motorway boxes' in London. A plan for a series of ring roads in and around London had long been mooted, and in the early 1960s, when Ernest Marples was the transport minister, a scheme for three concentric motorways was put forward. The inner one was to run in a circle three to four miles from Charing Cross, and it would have resulted in the demolition of 20,000 homes. As I quoted in my previous book, *Are Trams Socialist?*, this 'was seen as unproblematic by the British Road Federation [a pressure group for the road construction lobby] because, according to evidence it gave to the GLC, "much of the route lies in obsolete areas which urgently need rebuilding"'.[4] Those 'obsolete areas' were places such as Blackheath, Hampstead and Chelsea, which fortunately were saved when the scheme was scrapped in 1973 by a newly elected Labour Greater London Council, reversing the party's previous policy of supporting the initiative.

There were a couple more attempts to build roads in London, and while elsewhere the centre of many provincial towns and cities would yet be bulldozed to create space for traffic jams, the tide had turned. There was a recognition that not all transport problems could be solved by building extra facilities for motor vehicles and that public transport still had a key role to play. The oil price shocks of the 1970s, together with a growing realization of the greenhouse gas effect, began to put the motor manufacturing industry on the defensive.

It is not much of an exaggeration to say that the automotive industry has been in an almost permanent state of crisis since then. There has been a series of major upheavals, culminating in consolidations and factory closures across the world. The US industry consolidated to become dominated by four firms who expanded through acquisitions in Europe and beyond. The fallout from the 2008 financial crisis stimulated yet more consolidations and mergers. In the United Kingdom, by the end of the 1960s there were also just four major manufacturers. Three of these were owned by American companies and the fourth, British Leyland, soon had to be bailed out, effectively becoming a nationalized company in 1975. Competition from Japan, which eventually moved assembly plants to Britain, effectively killed off any long-term hopes of recovery. British Leyland, which became Austin Rover Group, was sold to British Aerospace in 1988, and was then bought by BMW six years later. This meant that for the first time in over a century there was no British-owned mass car manufacturer.

Despite the success of Nissan and Toyota in producing cars in Britain for export, the industry has continued to face significant challenges. Moving production to lower-cost countries has helped, but it has not solved the problem of selling sufficient numbers of vehicles at a profitable price.

The image of the industry has also taken a battering. It has fought rearguard actions against every environmental regulation, which has led to its reputation being almost as tarnished as that of the tobacco industry.

The story of lead in petrol illustrates this point well. Lead had been added to petrol for decades to prevent 'knocking', but in the 1960s environmentalists began to point out that this presented a major health risk. Lead in the air was being breathed in by children, affecting their brain development and causing other health and even behavioural issues. Millions of small children across the world suffered as a result. The removal of lead only happened as a result of a long campaign by activists, including, notably, Jill Runnette, who called herself a 'Wimbledon housewife' and who took on the oil giants, the government and scientists to convince the public of the danger of lead in petrol. The campaign was later spearheaded by Des Wilson, a former director of Shelter, who eventually won the battle. The motor vehicle manufacturers resisted at every stage. As Geoffrey Lean put it, writing in the *Independent on Sunday* in 1999 on the day that leaded petrol was finally banned in the UK: 'Oil companies have continued to make leaded petrol, long after its dangers have been accepted by doctors and governments, and substitutes became known.'[5]

The reluctance of the industry to address environmental issues as well as its readiness to use fair means and foul in their efforts to retain profitability were well illustrated by the revelation that amendments were made by Volkswagen to its cars' software in order to dishonestly pass emissions tests. The scandal, inevitably dubbed 'dieselgate', revealed that Volkswagen had systematically cheated on its emission ratings by dodging regulations aimed at reducing pollution from vehicles. The car manufacturer's cheating was uncovered by the US Environmental Protection Agency, and its scale and extent were utterly extraordinary. The agency found that Volkswagen had deliberately changed the way its engines functioned in order to activate emission controls only during laboratory testing and not when running normally. Therefore, while during testing the nitrogen oxide output met regulated standards, in normal running the emissions were up to forty times the legal limit. This was not some insignificant programme run by a rogue element within the company: it was a deliberate policy from the very top, and it involved a staggering 11 million cars.

While this scandal demonstrated the extent to which vehicle manufacturers will resist attempts to reduce the environmental damage caused by their products, it also marked a turning point as the industry recognized that it would have to change. It strengthened the hand of those who were already pushing for a far sharper focus on electric cars, as demonstrated by Volvo's plan to focus on the development of electric cars. It has promised that half the cars it produces in 2025 will be all-electric, and it has stated that it hopes to have produced one million electric cars by that date.

Globally – and particularly in Europe – however, there are still too many factories producing too many vehicles. According to the economist Professor David Bailey, in an interview with the author:

> There is still overcapacity in the car industry in Europe – perhaps still 20 per cent over supply. There is a reluctance among chief executives to close down factories as they do not want to lose out on capacity. Governments, too, tend to be reluctant to close down factories because of the political consequences.

Bailey argues that there needs to be a Europe-wide agreement, as there was with the steel industry a couple of decades ago, to agree a plan to cut capacity, but the auto industry is too fragmented.

For the time being, the industry remains the biggest manufacturing business in the world. After a slump following the financial crash, motor vehicle sales boomed in the United States, with annual sales hovering around the 17 million mark throughout the past decade – until the pandemic caused a collapse in demand. In the European Union, too, sales have recovered to around 16 million per year, but that is still some way short of the peak achieved in 2007, just before the financial crash, of 18 million. The emerging nations are not currently a happy hunting ground for the manufacturers, with only China showing significant growth, and even there the rate of sales is far lower than it was previously. Russia, hit by sanctions, and Brazil, affected by an economic downturn, had both shown large falls even before the pandemic.

These emerging nations were the great hope for expansion by manufacturers, who are now having to refocus their attention on their traditional markets in the United States and western Europe. And even here, storm clouds are gathering. In the mid 2010s, regulators in both the United States and Europe had mandated that by 2025, all new cars that are sold must, on average, achieve sixty miles to the gallon, compared with the current figure in the United States of eighteen miles to the gallon. In 2020, President Donald Trump reduced the requirement in the US to forty miles per gallon. Nevertheless, a step change increase in fuel efficiency will be needed, and this will cost manufacturers billions to develop. The cheapness of oil, regarded as positive for the industry, is a deterrent to improvement since it tempts purchasers, especially in the United States, to go for larger, gas-guzzling cars. Consequently, it is hardly surprising that the possible advent of these regulations is spurring manufacturers to look at different business models, a search which has been made all the more urgent by the collapse in the market as a result of the pandemic. Indeed, there is an element of desperation among manufacturers, who have faced an unprecedented collapse in demand.

This is the context for the drive towards electric and autonomous cars: the need for the manufacturers to keep making money and, even more dramatically, to stay in business, since they face an existential threat. They cannot admit, of course, that it is concern with their plight that is pushing them towards the autonomous revolution. Instead, they are presenting it as what customers want: a new type of car, and a different model of car ownership that is so much better than the current one.

As we shall see, there are many clear parallels between the hopes and aspirations that first stimulated mass car ownership and those now promised for the driverless age. This time, neither the public nor policy makers should be fooled. The risk is that we could be about to make the same mistakes that brought us the nightmarish aspects of our motorized world if we allow the outlandish claims of the advocates of autonomous cars to set the transport agenda of the future.

Chapter 2

The hard sell

Both the car manufacturers and the tech companies that are working on autonomous vehicles are trying to create a new reality in which the advent of driverless cars is inevitable. They never stop telling us it is going to be a wonderful world. A *different* world. A *better* world.

Both of these groups are old hands at selling visions of the future. As described in the previous chapter, cars were marketed as the great liberating force. It was that vision which the manufacturers cleverly invented and publicized to help build up consumer demand for their cars, ushering in the motoring age that dominated the past century. The manufacturers needed government support along the way – often financial but also political – to establish the right conditions and legislation. And so it is with the tech companies today: Silicon Valley has used many of the same gambits. You only have to spend a few minutes listening to tech pioneers such as the founders of Facebook or Airbnb to hear how their 'disruptive' companies are going to create a more inclusive and happier world. 'Don't be evil', Google used to say; that has now been changed to 'Do the right thing'. Silicon Valley built its very reputation on harnessing technology to make us happier, more fulfilled and more satisfied human beings. Technology, they argue, is the catalyst for a better life – and incidentally, don't get in our way as we are creating that better life, or put any constraints on our growth.

In pursuing the vision of a driverless future for cars, both these groups of companies present a vision that is invariably utopian. At Expo 2010, held in Shanghai, the pavilion hosted by General Motors and its local partner the Shanghai Automotive Industrial Corporation gave the public a glimpse of how they viewed the automotive world of 2030. Introducing it, the then-boss of General Motors Kevin Wale said: 'Our vision for the future is free from petroleum, free from emissions, free from accidents, free from congestion, and at the same time fun and fashionable.'[6] His PR people had clearly written the script because he eulogized further, but ungrammatically:

> It is a vision that will transform the lives [sic] around the world, bringing people and cultures closer together, a future in which people, motor vehicles and roads coexist in harmony and a future where motor vehicles no longer have a direct impact on the natural environment.

The motoring industry was, it seems, building this new world almost single-handedly, and it must have taken great effort from Wale to stop short of saying that peace on earth for all people was his ultimate goal.

In other words, Wale pitched the autonomous electric car as motoring with no downsides, with the motor manufacturers opening up the pathway to heaven. No expense was spared putting forward the concept to visitors to the Expo. In *The Great Race: The Global Quest for the Car of the Future,* Levi Tillemann describes how

> visitors [were] strapped into five-point harnesses as an IMAX-sized movie with computer generated imagery flew them through a bright, crisp virtual reality. Electric pods raced through the streets at breakneck speeds. Stoplights, traffic jams, and even drivers were gone.

All this, according to the two car manufacturers, would be achieved by 2030, when 'China would be animated by a living network of safe, efficient, zero-emission vehicles in constant communication with each other and the environment'.[7]

The film showed a blind girl racing through the canyons of Shanghai's tower blocks in her safe personal mobility pod while the conductor of the city's symphony orchestra, freed of driving obligations, was reviewing his scores, and a pregnant mother was being rushed to hospital just in time by an autonomous ambulance (we will return to the subject of emergency vehicles later). After watching the film, the spectators were treated to a view of real life electric vehicles like those they had watched on screen, trundling autonomously round the building. The vision, as Tillemann wrote, was tantalizingly close, except, as he pointed out, the visitors re-emerged into the exhaust-laden smog swamps of Shanghai, where 'real life meant navigating manic waves of oil-burning SAICs, VWs, Audis and Buicks'.[8] We are at the midpoint between the expression of this vision and its supposed implementation, and yet there is no sign that any of the requirements needed to bring it about are even at the first stage of development.

If autonomous car heaven is the carrot, the stick is safety. The propagandists for autonomous vehicle technology invariably start their presentations with accounts of the terrible death toll on the roads. And terrible it certainly is, especially in the United States where the main drive for autonomy is coming from. Moreover, the toll is increasing, thanks to cheaper fuel prices – which leads to increased traffic – and a failure to clamp down on people using mobile phones while driving. Google's main selling point for the driverless car is safety. The mission statement of Waymo, the name now being used for Google's autonomous car project, is: 'We are a self-driving technology company with a mission to make it safe and easy for people and things to move around.'

Figure 1. Google's driverless car.
(Photo by Grendelkhan. *Note*: full photo credits
for all figures appear on pages 145 and 146.)

In 2016, Google's then head of the project, Chris Urmson, elaborated the company's plans to a congressional committee and started off by stressing the (undoubtedly quite remarkable) death toll on the roads of 36,120 in 2019 (a reduction on recent peaks, during which the total exceeded 40,000). He pointed out that '94 per cent of [the annual 6 million] accidents in the US are due to human error'. The implication of his statement is that driverless cars would, by contrast, have no accidents:

> When Google started working on self-driving vehicles over seven years ago, our goal was to transform mobility by making it safer, easier, and more enjoyable to get

around. What drives our team is the potential that this technology has to make our roads safer.[9]

Another carrot is to increase access to vehicles for people with disabilities, like the blind girl in the Shanghai film. This is frequently referred to by the promoters of autonomous vehicle technology and one can almost hear the violins in the background as people are rescued from a life of immobility by the new technology. Urmson explained how the technology would allow disabled people to work. He cited the case of 'Justin Harford, a man who is legally blind', who he had met at a recent seminar he had attended: 'Justin said: "what this is really about is who gets to access transportation and commerce and who doesn't and I'm frankly tired of people with disabilities not being able to access commerce"'. This is, of course, far more relevant in the United States, where car dependency, apart from in a few major cities, is almost universal given the paucity of public transport. Nevertheless, similar arguments are being deployed by advocates of the technology in the United Kingdom too.

The disability lobby, at least in the United States, seems to have been taken in. In his written evidence to a congressional committee, Clyde Terry, the chair of the National Council on Disability, could not disguise his enthusiasm:

Aside from being one of the most exciting innovations in transportation since the Model T began rolling off the assembly line in 1913, [autonomous vehicle] technology holds tremendous promise for many people with disabilities and seniors who currently lack access to independent transportation. ... Autonomous vehicles hold great promise to advance social inclusion by offering people with disabilities independent mobility to get to school, jobs, and all places that Americans go each day. They also offer the possibility of ending the isolation that many people who are aging experience by keeping them connected with others and to activities that are often lost when we lose the ability to drive.[10]

In a letter to the same committee, Parnell Diggs, the Director of Government Affairs for the National Federation of the Blind, went even further:

We anxiously anticipate the day that all blind people will have the opportunity to drive independently, and we believe that autonomous vehicles will make this day possible.

Urmson also mentioned a 'woman in Southern California who lost her ability to drive 15 years ago [who] told us, "my life has become very expensive, complicated, and restricted" since she had to start paying drivers and enduring long waits for buses and trains'. This is revealing. Urmson clearly envisages a world where autonomous vehicles replace public transport. He does so more explicitly later in his statement: 'The technology also has the potential to reduce current Federal spending pressures for roadways, parking, and *public transit*' (my emphasis).

The more one digs into the future envisaged by this new world of autonomy, the more it becomes clear that driverless vehicles are seen as a replacement for not just cars but other forms of transport too. Moreover, as we will see in the next chapter, the concept is predicated on a new form of ownership – or rather lack of ownership. Transport, it seems, will be provided by driverless pods that are not owned individually and will replace cars, taxis, buses and quite possibly urban railways and trams. Who would bother walking to a bus stop or station if they could whistle up a pod that would drive them right to their destination instead? Arunprasad Nandakumar, a specialist on autonomous cars for the investment company Frost & Sullivan, made the point more explicitly in April 2017 when commenting on the government's decision to invest £38 million in driverless car technology:

> Considering recent reports suggesting the average driver in London loses over 100 hours a year in traffic, 25 percent more than any other city in Europe, the focus on

autonomous shared services designed for London will likely help in better balancing the heavy reliance on public transit services in the city.

In other words, by making road traffic more efficient, people will be attracted off public transport into their own driverless cars.

The hype put out by the tech and auto companies has found a ready audience in the media. The idea of driverless vehicles has been accepted uncritically in nearly all reports on their progress, and every statement about them is recycled, with extra zest, by both conventional and social media. The headlines, however, reflect the hype, which is sometimes debunked by a careful reading of the articles. Take the *Daily Telegraph*'s coverage of Uber's 'driverless taxi service', announced in September 2016:

> In an ambitious experiment, a fleet of cars laden with lasers, cameras and other sensors – but with no one's hands on the wheel – have been deployed by the web-based ride service on the challenging roads of Pittsburgh, Pennsylvania, steering themselves to pick up regular Uber passengers who are used to being fetched by cars driven by humans.[11]

The story spread across the world. Under the headline 'Uber launches driverless car service in Pittsburgh', an article in the *Toronto Star* began: 'A fleet of self-driving

Ford Fusions are giving Uber riders in the city the chance to get a glimpse of the future'.[12] But it wasn't much of a future since the cars were not self-driving at all. As was revealed later in the newspaper stories, they actually had not just one but two people in them: a test driver and an engineer monitoring their performance. It had, in fact, taken Uber two years to reach this stage since the cars' computers had to be loaded with very detailed maps of the city and even then the test drivers proved to be very busy people, as we will see in the next chapter.

Figure 2.   Uber's self-driving car test driving in downtown
San Francisco. (Photo by Diablanco.)

Ultimately, the Uber experience was to be an unhappy one for Pittsburgh. Within six months of the introduction of the scheme, the city's politicians and Uber executives were engaged in a war of words. The mayor, Bill Peduto, had hoped that Uber would help the city lever in a $50 million grant to help revamp its ailing public transport system, but Uber refused to play ball, confining its goodwill to a $10,000 donation to a local women's centre. Promised jobs did not materialize and local protesters, representing bus drivers and passengers, started a #deleteuber campaign after the company continued to provide a service to the airport despite other taxi drivers refusing to do so in response to President Trump's anti-Muslim travel ban. Peduto had originally been delighted at Uber's arrival, with a well-publicized inaugural journey in the first 'driverless' taxi. However, the relationship quickly soured when he found himself being charged for a ride in one of the taxis, despite Uber's promise that all journeys would be free during the experimental phase. The mayor also claimed that the company broke its commitment to allow city officials access to the data from trips taken by its passengers. As an embittered Peduto told the *New York Times*: 'When it came to what Uber and what Travis Kalanick wanted, Pittsburgh delivered, but when it came to our vision of how this industry could enhance people, planet and place, that message fell on deaf ears.'[13]

Waymo's 'driverless taxi' service around Phoenix, Arizona, which launched in October 2019, more than a year later than originally announced, has not fared much better. First, it is not really 'driverless'. Every vehicle is monitored remotely by a controller who is able to intervene if a clear error is being made by the vehicle. There is also a back-up van with a spare driver in it, so that any stranded vehicle can be rescued. Such is the concern of Waymo about bad publicity that it both pre-selects its passengers and has them sign non-disclosure agreements before they are allowed to use the service.

Additionally, there have been numerous problems with the vehicles. They drive very conservatively, and local residents have made numerous complaints about the fact that they slow down the traffic. For instance, they automatically stop for three seconds at every stop sign, far more than a human driver needs to work out if it is safe to proceed. The vehicles appear to have a difficulty with turning left – across the path of oncoming traffic, in other words – particularly when there are two lanes of traffic turning in this way. They also have difficulty recognizing traffic lights controlling the flow of traffic onto highways.

The vehicles are confined to a geofenced area of suburban Phoenix, but interestingly they are allowed out at night, though not when it's raining or during Phoenix's frequent dust storms (known to locals as 'haboobs'). One serious problem is that the cars (most are in fact minivans) find it difficult to discern individuals, whether they are pedestrians or cyclists in a group, and they find it particularly hard to work out what their intentions are. They therefore often come to a sudden halt when they first spot such groups. They also find stopped cars

– such as those turning into a shopping mall or ones that are stationary because of an accident – very confusing. Because they are programmed not to break the law, they are unable to take rapid evasive action in such circumstances, so they simply come to a halt. They have also failed to master situations on narrow roads where oncoming traffic may be on the wrong side of the dividing line. A reporter who spent time in Phoenix talking to local people about Waymo's vehicles found that there was one three-word phrase that repeatedly came up when discussing them: 'I hate them'.

This contrast between the imminent driverless revolution presented in the headlines and the reality in the details of the story is an almost daily feature of news coverage of the issue. On the day I was working on this chapter, my eye was caught by a headline on the website of the *Daily Mirror* that read 'Domino's launches ROBOT pizza deliveries in Europe'. The article continued: 'The pizza delivery company is testing out a novel way of carrying out deliveries.'[14] This, again, had become a global story, and a website called TechCrunch was typical, illustrating it with a picture of a young woman picking up a pizza from a 'self-driving' car.

The truth, however, was again far more banal. The 'self-driving' cars will have drivers but they will stay in the car 'behind darkened windows' while delivering the pizzas. TechCrunch revealed that the point of the experiment, run jointly by Domino's and Ford, was not to test the technology but 'to see how people react to receiving pizzas via self-driving vehicles'. I'll short cut the process and save them the bother of continuing the experiment: I would guess that people living on the ninth floor of a New York apartment block or ordering a pizza so they can settle down to watch a football game would not be pleased at having to rush out to pick up their pizza from a car that might be parked far down the street, especially if it were raining or snowing.

Actually, there is a serious point here. This kind of experiment is clearly trying to produce benefits – such as cutting the cost of employing drivers – for the companies involved, with a concomitant deterioration in service for their consumers. Sure, there may be a small reduction in the cost of delivering the pizzas, which could be passed on to the customer, but only for providing a far worse service.

These articles lead to an informal competition to produce the most sensational story. At the time of writing, my favourite was from *Forbes Magazine*,[15] an authoritative US business fortnightly with a circulation approaching a million, in which David Galland, a partner and managing editor of Garret/Galland Research, wrote: 'my research leads me to believe that there will be 10 million self-driving cars on the road by 2020, with one in four cars being self-driving by 2030'. He adds: 'But I think those estimates (especially the one for 2030) are too conservative.' He based his prediction not on any assessment of the current state of the technology but on the rate of adoption of other innovations:

> The technology adoption cycle has been steadily compressing. While it took approximately 50 years for electricity to be adopted by 60% of US households, it took cell phones only about 10 years and ... smartphones only about five years to reach the same penetration.

Needless to say, there were no autonomous vehicles on the road by 2020. A more sober and accurate assessment was made by the *New York Times* in August 2017 in an article responding to the claim by Travis Kalanick (by then recently deposed as Uber CEO) that its vehicles would be entirely driverless by 2030:

> Most experts (including those previously bullish on self-driving technology, such as the editors of *The Economist* magazine) have recognized that autonomous vehicles are at least 20 years from fruition. We will continue to see various experiments, and autonomous service vehicles used in very limited settings, but Mr. Kalanick's promise of a self-driving transportation grid dominated by Uber is pure science fiction.[16]

Galland was guilty of one of two things: either his research was so cursory that he had not properly examined the progress of the technology, or he was prepared to overlook his findings in order to contribute to the hype that surrounds this industry. An entry on his website at the time suggested we would soon 'be receiving our morning coffee by flying robot'.

Indeed, much of this type of coverage is generated by reports from companies with a vested interest. For instance, a report produced by the Transport Research Laboratory (TRL) carried the headline 'Four out of five people think autonomous vehicles are a good idea'. However, as TRL's chief scientist and 'Research Director, Transportation' Alan Stevens admitted, the report was based 'on a relatively small and self-selecting sample of people'. Indeed, the sample was 233 people who, far from being randomly chosen, had answered an online survey. In scientific terms – which someone as eminent as Mr Stevens ought to understand – the data were therefore biased. For all we know, every respondent could have been a client of TRL or one of Mr Stevens's mates. Indeed, the report's introduction says: 'A large proportion (64 per cent) of the respondents were recruited via their registered interest in a major UK trial of automated vehicles, the GATEway project'. So people who are interested in autonomous vehicles think they are a good idea? Gosh. In the press release on the report, TRL, once government owned but now in the private sector, boasts about its extensive involvement in the sector: 'TRL is at the forefront of smart mobility with a current portfolio of connected and automated vehicle projects in excess of £50m including projects such as GATEway, MOVE_UK, Driven & Streetwise'.

Galland clearly also wants to create the right atmosphere for further funding of schemes, and his *Forbes* article goes on to list the numerous advantages of autonomy. He repeats the safety predictions, assumes that autonomous cars will allow older and disabled people to access cars, and suggests that autonomy will free up large swathes of car parking space in city centres. He points out that without drivers, military vehicles will be able to enter far more dangerous places, and he claims that we can 'say goodbye to traffic congestion'. Many of Galland's arguments are unsubstantiated nonsense or wild speculation, as will be demonstrated in later chapters. However, the appearance of such an article in a serious business magazine demonstrates the extent to which the hype put out by the tech and auto companies – together with exaggerated media coverage and support from consultants and advisers – is creating an autonomous car bubble, not unlike the dotcom bubble of a decade ago.

This optimistic presentation of the driverless car revolution in the media is vital in creating the climate that the tech and auto companies hope will ensure that the regulatory and legislative changes they are seeking can be realized. Readers tend to remember stories' headlines, rather than their detail, which makes the misleading media coverage potentially very influential. There is a widespread perception that autonomous cars are already operating and that the technology is virtually ready. I have personally had to explain countless times to friends and acquaintances that there are no autonomous vehicles operating unaided on public roads anywhere in the world without a test 'driver' aboard who is able to take control, and that tests and pilot schemes are currently confined to specific routes or areas determined by the software available in the vehicle. As we have seen, even Waymo's much-publicized 'robotaxis' are limited by strict geofencing, only operate under the right weather conditions, and are remotely monitored.

Thanks to the constant drip-feed of these stories, a casual assumption is growing among the public that the technology is ready to be introduced in the near future, with no recognition or analysis of the problems highlighted in subsequent chapters in this book. The predictions from the manufacturers, tech companies, various industry bodies and the media pour out. There was even a website[17] dedicated to listing published predictions, but it has now gone out of business. For instance, no less a figure than Google founder Sergey Brin got it badly wrong when he said in a speech to mark the signing of California's autonomous car legislation in October 2012 that driverless cars would be available to Google's employees within a year and that they would be on the market commercially in 'no more than six years', i.e. by 2018. Neither of these predictions turned out to be correct, so why should we believe Google's latest claims?

Ford's chief executive Mark Fields said in February 2015 that fully autonomous cars would be on the market by 2020, though he did not say they would be made by Ford. Anthony Foxx, then US Transportation Secretary, claimed in January 2016 that driverless cars would be in use all over the world by 2025. And yet even an article headed 'Elon Musk is right: driverless cars will arrive by 2021' on a website called 'The Next Web' concluded that '2021 might be a tad optimistic but it seems we are closer than decades away'.

It is not only politicians, the auto manufacturers and tech companies making these predictions either. Uber's (later-ejected) chief executive Travis Kalanick tweeted in August 2015 that he expected Uber's fleet to be driverless by 2030 and that the service would then be so inexpensive and ubiquitous that car ownership would be obsolete. However, as we have seen, Uber has not found it so easy to dispense with drivers. Uber, in fact, sees getting rid of drivers as key to its ultimate profitability; it is currently losing billions of dollars annually, more than any other tech start-up company: a staggering $8.5 billion in 2019. And it will have suffered from the pandemic, too, as many people were prevented from using its vehicles. Uber is betting the house on a particular model of car use in the future, and so is Google. They envisage a world where autonomous electric cars are not owned individually but are instead a fleet of taxis that can be called up instantly, used for a particular journey, and then sent on their way. It requires a triple revolution: cars need to become electric powered, autonomous and pooled – and possibly connected.

Chapter 3

The triple revolution

One of Google's key selling points for autonomous cars is the idea that parking space will be liberated. In US cities, thousands of acres of parking lots would become available if people no longer needed to park their cars near their place of work or the shops they are using. There is also the promise that swathes of suburban streets could be grassed over since they would no longer need to accommodate parked cars. In the evidence he gave to the congressional committee mentioned in the previous chapter, Google's Chris Urmson stated that in the United States parking takes up an area the size of Connecticut and he implied that this space would be liberated by the advent of autonomous cars.

Like many of the ideas posited by the autonomous car lobby, this vision would require a complete rethink about the way that we own and use cars. Let's look in turn at the progress made in each of the three areas of the triple revolution.

Electric

It is inconceivable that shared-use autonomous cars would be powered by internal combustion engines. There are numerous reasons for this. Electric vehicles are far more reliable and easier to refuel, as they do not require a visit to a service station. They are emission free at the point of use (though ultimately their greenness depends on the source of the electricity) and are cheaper to operate (though more expensive to purchase, for the time being at least). Maintenance costs are also far lower for electric vehicles. All this is backed up by the fact that the car manufacturers recognize that the days of petrol- and diesel-powered vehicles are numbered. By 2020, fourteen countries across the world had announced end dates for their production, including Britain (scheduled date 2035) and France and Canada (2040), while China, despite announcing its intention do so, has not yet fixed a date.

The big problem with electric cars has been their limited range, compounded by a fear of running out of power. This means that even if the range is stated as 100 miles, drivers are unlikely to dare to do more than, say, 80 miles, for fear of finding themselves stranded, although newer models now have very sophisticated meters showing precisely how many miles remain until the battery runs out. This fear is in part because of the lack of charging facilities, which are currently found in very few locations. Moreover, getting stuck is not a matter of spending five minutes filling a tank but of waiting for perhaps an hour or so for the batteries to charge sufficiently to get back home – and of course the driver cannot trudge off to the nearest charging point, as one can with a petrol can, but instead has to get the car to that charging point. Added difficulties include the fact that there are numerous different charging mechanisms that are not compatible with one another, and that routine charging takes four to six hours, which effectively means it has to be carried out overnight or on a day when the car is not being used.

The industry is working flat out to remedy these issues, with several holy grails: increasing the range of the cars, decreasing their weight to power ratio, reducing vehicle purchase costs, improving battery technology and using sustainable sources of energy. It is really not until all of these are achieved that the mass replacement of conventional vehicles with electric ones will become feasible. One much greater issue is the fact that it is difficult for many people – those living in, say, high-rise flats or terraces, where it is impossible for residents to park outside their own front door – to run a cable over the pavement from their house to their car. The network of charging points required to service such housing would be incredibly expensive to provide, and it is not at all clear who would pay for it.

Because of the vehicles' shortcomings, their greater cost and the lack of charging points, take-up has been slow, though it has accelerated recently. At the high end, Tesla – a start-up firm created by Elon Musk, the serial entrepreneur and inventor, as well as a powerful and leading advocate for driverless vehicles – produced its Model S in 2012. It has a range of 200 miles for the basic model, extending to 355 for the most expensive version. Tesla does not use single-purpose, large battery cells like those in other electric vehicles. Instead, it uses thousands of small, cylindrical, lithium-ion cells similar to those in laptops and other electronic devices. Tesla has focused on high-end vehicles, typically in the £50,000–£70,000 range, and has struggled to provide the cheaper models that would turn it into a mass-market manufacturer. Globally, the company reached the 1 million sales mark only in March 2020, but it still has ambitions to expand rapidly despite never having turned a profit in any full year of business.

At the other end of the market, the Nissan Leaf, the best-selling all-electric vehicle, is a hatchback aimed at the low end of the market, with a range of just over 100 miles in its more expensive version. Nearly 500,000 had been sold worldwide by the spring of 2020, but at just under £30,000 they are still far more expensive than their petrol or diesel equivalents – though that difference will be offset over time by fuel savings.

Also at the low end, China has produced several small cars that have a maximum speed of 50 miles per hour and ranges of around 100–120 miles that are designed to be used only in cities, as they are too slow for highways. The market is growing, thanks partly to support from the Chinese government, which is anxious to reduce pollution in cities (it has already banned petrol and diesel two wheelers in urban areas). Annual sales in China reached 1.2 million in 2019, if hybrid vehicles are counted as well as plug-in ones, but that still represents only around 5 per cent of the new-vehicle market.

Electric car sales amounted to around 2 million, globally, in 2019, out of a total of 65 million passenger cars sold. This shows that they remain a niche market. While sales of electric vehicles are clearly set to grow, there are major logistical and practical problems to overcome before they can become the car of choice for most people.

The key issue is whether enough batteries can be produced. Tesla is constructing what will become the biggest building in the world (in terms of its footprint) in Nevada to manufacture lithium-ion batteries, and it is eventually expected to produce enough batteries for 1.5 million cars per year. Elon Musk is planning several more such 'gigafactories', but there is clearly, at the moment, huge undercapacity of battery manufacturing in relation to the demand that would be created if even 10 or 20 per cent of cars, let alone a majority, were electric powered. While this is not an insuperable problem, there are also questions about the availability of sufficient lithium to produce these batteries. An article in the *Financial Times* in June 2017 warned that prices might begin to rise rapidly as production of electric vehicles increases, and it suggested that there would need to be a twenty-fold increase in worldwide production of lithium to meet demand by 2030 in order to 'electrify the world's fleet of vehicles'.[18]

Autonomy

There is much confusion, which is often reflected in media coverage – or even originated in it – on the definition of a 'driverless' car. In this book I have used the more accurate expression 'autonomous' as this is less binary, offering a range of possible levels of technology. The Society of Automotive Engineers in the United States has drafted a set of definitions, ranging from 0 to 5, to measure the level of autonomy, and their classification has been recognized by the Department for Transport in the United Kingdom and adopted by the industry worldwide.[19] The figure that appears on the facing page details the various levels.

Level 0 describes today's basic vehicles. The driver is responsible for practically everything: steering, signalling, braking, acceleration, observing traffic conditions and reacting to them, and so on. Cars that are fitted with basic driver aids, such as cruise control or reversing sensors, are classed as **Level 1.** In effect, neither of these levels has any real autonomy.

L0	Driver only	• Driver in complete control • No intelligent system aids • Example: N/A
L1	Assisted ('hands on')	• Driver must control all normal driving tasks (steering, acceleration, braking, monitoring the environment) • Discrete and limited driving tasks can be selected to be carried out autonomously • Example: parking assistance
L2	Semi-automation ('hands off')	• Driver must monitor the system and environment continuously • The system is allowed by the driver to control steering, acceleration and braking when the driver determines specific circumstances are suitable • Example: traffic jam assist
L3	Conditional automation ('eyes off')	• Driver must be in a position to resume control, but need not monitor the system and environment continuously • System steers, accelerates and brakes in circumstances deemed suitable by the driver; the system knows its limits and will request driver assistance with sufficient warning • Example: highway patrol
L4	High automation ('mind off')	• In situation deemed suitable by the driver, system has full control; no input required by the driver • System steers, accelerates and brakes in complete control in this defined situation • Example: urban driving
L5	Full automation ('steering wheel optional')	• No driver required • System performs all driving tasks in any environment or situation for the entire journey • Example: on-demand 'taxi' service

Level 2 vehicles are rather more sophisticated as they are fitted with devices such as wide-angle cameras, GPS sensors and short-range radar. As a result, they can adapt their speed to the surrounding traffic automatically, maintain a safe distance from the vehicle ahead, remain within their own lane, and some can park themselves. The vehicle can operate itself, with no input from the driver either through the pedals or the steering wheel, for short periods of time, but the driver must be ready to take control of the vehicle instantly.

This level was called 'Autopilot' by Tesla, but an accident in Florida in 2016 called into question that terminology. Joshua Brown – a strong advocate for Tesla cars who had produced several clips for YouTube about his 'Tessie' – had put his car on Autopilot when the sensors failed to pick up the fact that a large white truck was crossing its path. The car steered itself under the truck, killing Brown instantly. The driver of the truck said that when he approached the vehicle a Harry Potter film was playing, but the Brown family lawyer denied that he was watching the screen at the time of the accident. As a result of the crash, Tesla amended its Autopilot function so that drivers cannot repeatedly ignore warnings to take back control of the vehicle, as Brown had done on seven occasions in the half hour before the fatal crash. However, Tesla denied any responsibility for the accident because it 'does not allow the driver to abdicate responsibility'. A report into the accident by the National Transportation Safety

Board found that Brown had barely touched the controls during the thirty-seven minutes prior to the crash and had set the cruise control to 74 miles per hour, above the 65 miles per hour speed limit. Elon Musk, the boss of Tesla, accepted that the car should have spotted the truck and that updated versions of Autopilot would do so, but he argued that Tesla was not responsible and he resisted any attempt to slow down the introduction of new technology, suggesting that 'the perfect is the enemy of the good'.[20] Musk argues that since new technology will save lives, introducing it, even if it is not perfect, still leads to an improvement in overall safety and consequently reduces the death toll on the road.

Despite Musk's confidence, the problems posed by Level 2 are exacerbated in **Level 3**, in which the main difference is that while the driver must still remain vigilant and ready to intervene in an emergency, responsibility for all the critical safety functions is shifted to the car. The added risk is that drivers will lose focus, and therefore not know when or whether to intervene, or they will be too slow to react.

An even greater conundrum about the long-term effect of Level 3 is the deskilling of motorists who become accustomed to adopting a solely supervisory role when driving. Driver aids in general have already been shown to reduce the skill set of motorists. According to Adrian Lund, president of the Insurance Institute for Highway Safety in the United States:

> There are lots of concerns about people checking out and
> we are trying to monitor that now. Everything we do that
> makes the driving task a little easier means that people
> are going to pay a little bit less attention when they're
> driving.[21]

This is a well-known phenomenon known as the 'paradox of automation'. Experience of other industries, notably nuclear power, suggests that there are enormous risks with control systems that relegate the operator to a managerial role whose only job is to intercede in the case of an emergency. In both the aviation industry and the maritime industry there have been examples of the negative effect of this deskilling. In aviation, pilots who have become used to relying on autopilot, which flies the aircraft much of the time, have found it difficult to react correctly to emergency situations. This was most apparent in the disaster involving an Air France A330 that plunged into the Atlantic in June 2009, killing all 228 people on board. Hitting an area of turbulence, the autopilot disconnected probably as a result of ice forming on inlet tubes, and the pilots subsequently misread their instruments, reacting incorrectly by making the aircraft climb when they thought it was descending. That caused a fatal stall from which the pilots were unable to recover. The crash report found that the accident was a result of the pilots' failure to assess the very unusual situation correctly. Indeed, as Tim Harford says in his book *Messy*, 'the better the automatic system, the more out-of-practice humans will be and the more unusual will be the situations they face'.[22] And the A330 had particularly good automation systems, too.

There have been other instances of overreliance on automation, such as the Asiana Airlines crash at San Francisco in July 2013. That was the result of the captain being insufficiently prepared to make a visual landing approach after the autopilot was disengaged because the wrong setting for the final approach had been selected. The aircraft landed in the Bay, short of the runway, but fortunately only three of the 307 people on board were killed. The accident report again pointed to the failure of the pilot to react to an unusual situation and blamed 'over-reliance on automation and lack of systems understanding by the pilots'.

While this over dependence on autopilots is a well-established and widely known risk in aviation, similar concerns have been raised more recently following shipping accidents. In 2017 two major accidents occurred between naval ships and large merchant carriers in busy shipping lanes in Asia. In both cases the merchant ships, which are now very lightly crewed, were using autopilot – one of these ships even continued on its course without slowing down or deviating for thirty minutes after the accident. As an August 2017 article analysing these incidents on the website 'The Conversation' says, 'There's a fundamental problem with the industry's reliance on technology to save the day when collisions become imminent, often in complex environments.' The article argues that the autonomous-car industry must learn from mistakes made in aviation that have led to disasters, such as those mentioned above, which, the author says, partly result from the fact that the technology has made pilots' tasks more difficult and complex, not easier:

> The airline industry trend towards higher levels of autonomy created new opportunities for confusion and mistakes – a situation called an 'automation surprise'. In another irony of automation, this cognitive dissonance often occurred in exactly the kind of unusual situation where advanced technology could have proven most valuable to their human operator. Yet, instead, they were doubly-burdened to sort through a confusing, dangerous and potentially escalating situation.[23]

Harford sums it up succinctly: 'Automation will routinely tidy up ordinary messes but occasionally cause an extraordinary mess.'[24]

Paul Jennings, Professor of Energy and Electrical Systems at WMG, University of Warwick, who is heading a team developing a simulator for driverless cars, is a great advocate of the technology, but he is particularly concerned about the implications of Level 3:

I don't like Level 3. The problem is the handover, and when do you take over, how you take over and in what circumstances. How quickly can people react? There are so many difficult questions that in a way it is easier to go straight to Level 4, though technically it is more challenging. In Level 4, it is clearer because the car is in control.[25]

Similar doubts were powerfully expressed by a House of Lords committee:

CAV [Connected and Autonomous Vehicles] could have negative implications for drivers' competence, making drivers complacent and overly reliant on technology. This is of particular concern in emergency situations, where a driver may react slowly to taking back control of a vehicle. It may be the case that for Level 3 vehicles the risks will be too great to tolerate.[26]

Some car manufacturers – including Ford, notably – are also so concerned about the issue of what are effectively semi-autonomous vehicles that they are skipping Level 3 altogether and going straight to **Level 4**, which involves cars being able to drive themselves in all circumstances while still retaining control pedals and a steering wheel, even though they recognize that this will take far longer. Ford found that its test drivers were dozing off at the wheel because, according to Ford's chief technology officer Raj Nair, 'it's human nature that you start trusting the vehicle more and more and that you feel you don't need to be paying attention'.[27] They tried a series of measures to keep the engineers awake, such as bells, buzzers, warning lights, vibrating seats and even shaking steering wheels, and eventually they put in a second driver, but it was to no avail as the drivers still either dozed off because of the smooth ride or failed to notice when they were in danger. Consequently, Ford decided not to proceed with the development of these semi-autonomous vehicles.

A similar line of thinking led to a major change in Google's approach to developing the technology. Originally, the company had adapted existing vehicles, adding features to make them autonomous but effectively only on what Americans call 'highways' and the British would term 'A roads'. The cars were driven manually to a highway and then put into autonomous mode. But from 2014 Google's Waymo comprehensively refocused its programme. Instead, the new cars would not only look completely different, so that they could better accommodate the huge range of sensors and radar equipment required, but they might also have no steering wheel or pedals, and passengers would only have the ability to override the computer in order make an emergency stop.

In his congressional evidence that I referred to in the previous chapter, Chris Urmson revealed that Google understood that the benefits of the technology, in terms of safety and access for disabled people, would only appear when the technology reached Level 4:

> In 2013, we decided that to fully realize the safety promise of this technology and serve the most people – even those without a license – our technology needed to be capable of doing all the driving, without human intervention [being] necessary. NHTSA defines this as 'fully autonomous vehicles,' or 'Level 4' on a NHTSA scale for automation established in 2013. Developing a car that can shoulder the entire burden of driving is crucial to safety: we saw in our own testing that the human drivers can't always be trusted to dip in and out of the task of driving when the car is encouraging them to sit back and relax.[28]

It is notable that Google, with all its money and big brains, had not understood until four years after its project had started that there is a problem with the concept of semi-autonomy.

Ford and Waymo argue that Level 4 technology should be safer. Level 4 vehicles are able to carry out all the driving functions by themselves, from the start of a journey to its end. The only proviso is that they must be restricted to roads for which mapping information has been loaded into the vehicle's computers. The vehicles would have to be confined – or 'geofenced' – to designated zones and prevented from straying outside them.

Level 5 is essentially the same as Level 4 but without any geofencing: it is autonomous car nirvana. Passengers will have no access to any controls apart from (possibly) an emergency stop button, so the vehicles will have to cope with every conceivable situation. They must have universal access to all roads, be able to travel on private or unmade roads, deal with all weather conditions, recognize the difference between a football kicked into a road and the child chasing it, negotiate through streets crowded with people, make the right decision every time at intersections, and, in short, cope with unpredictable human behaviour of every kind. To summarize, then, they must be capable of going anywhere, in any conceivable conditions, and they must be able to cope with the most unpredictable situations. That means travelling on dirt tracks off the map, in blizzards, thunderstorms or pitch darkness, with animals bursting out of bushes, children chasing runaway balls and crazy people doing crazy things. And lots more besides.

Then there is the reliability of automation. GPS, for example, has led people to drive into the sea and into lakes or to go straight over at T-junctions, and it has led HGVs onto narrow farm tracks and even down a set of stairs. And those are just the incidents that have been reported.

There is a great deal of confusion about the use of the terms 'driverless' and 'autonomous', which motor manufacturers and tech companies are wont to bandy about without a proper explanation of their meaning. They rarely refer to these levels, and this results in ill-informed journalists simply parroting the sort of claims made about the imminent arrival of driverless cars mentioned in the previous chapter.

Figure 3. Satellite navigation errors are not always so funny.
(Photo by David Stowell.)

Shared use

There is undoubtedly a trend, particularly among millennials, of being less interested in owning – or even driving – a car compared with the previous generation. A PriceWaterhouseCoopers report on connected cars suggests that 'urban residents in Western markets appear to be losing interest in owning their own cars, a trend exacerbated by their desire to move to urban areas, where cars simply aren't a requirement, and where public transport and ride-sharing apps can easily fulfil their needs'.[29]

Indeed. But this has to be put into context: global figures for car sales are still rising, in line with economic growth, and the shared-use concept has therefore only been adopted by a small urban-living minority. The idea that people will readily opt for communal vehicles is also questionable. Car clubs such as Zipcar have had some success. London has around 200,000 car-club members, a figure that has not changed much for several years, although Transport for London wants to see it rise to a million by 2025. Members have access to 3,000 vehicles, according to the annual survey of car-club use; there are 25,000 fewer independently owned vehicles on London's streets as a result. But this shows that it remains a minority interest. Zipcar reckons this figure could be tripled by the end of the decade, but its general manager in the United Kingdom, Jonathan Hampson, cautions against assuming that we will all be driving communal cars one day:

> Car clubs, though, are not for everyone and there are many people who still aspire to car ownership, even Millennials. I don't see a time when all cars will be shared.[30]

The AA's Edmund King describes the presumption of sharing as 'vastly hyped'. The flexibility and the door-to-door service offered by car ownership are two aspects that are difficult to dislodge from the public's mind.

There are good reasons for this. People like their own space, and driving in their own car, with the seat correctly adjusted and the radio tuned to their favourite station has its advantages. Many people want to keep their golf clubs or their favourite teddy bear in the car, given how much time they spend in it, and they would be loath to swap it for an anonymous pod. There is no reason to assume, either, that this aspiration will disappear with automation. Moreover, surprisingly, many people enjoy driving, as witnessed by the number of top-of-the-range models still sold, and they will not be satisfied with a boring shared vehicle, let alone one they cannot control. There is still a large minority of car owners who consider having a flashy vehicle on their drive to be a sign of their manhood – a fact reflected in much of the advertising by manufacturers.

There are numerous other points of resistance to the shared concept. The vehicles are unlikely to be cleaned between uses, particularly at peak times, and there will be concerns about availability for those dependent on travelling by road rather than public transport. The whole idea is very urban. It might work in dense central areas but it is unlikely to be attractive to people who live in sparsely populated suburban or rural housing, where shared cars will simply not be available without a considerable time delay. The local village's shared car (or cars) might have been nabbed by your neighbour, and then what? Whistling up a car on your app if you live a couple of miles off the main road, which is itself five miles from the nearest urban centre, will not be instant. Not like hopping into the car that is on your front drive. Having a fleet of vehicles sufficiently large to guarantee rapid availability will be expensive, and the business model for the concept is so far unclear, as discussed later.

It is difficult to work out the process by which people will begin to use autonomous vehicles summoned by smartphone so easily that they will be happy to stop buying cars altogether. These autonomous cars will have to be fleets owned by commercial companies, but who will take that risk, particularly with a technology that is so leading edge? This is not a Santander bike hire scheme or even its dockless rivals that do not require docking stations. Rather, it requires a complete change in our behaviour, and, as mentioned previously, there are many doubts about whether it is feasible at all. Moreover, as discussed in chapter 6, shared use is not dependent on autonomy but rather is determined by transport policies.

Interconnectivity

There is also a fourth potential revolution, which is that every vehicle on the road will be able to be in constant communication with every other vehicle, as well as with services providing information on traffic conditions, weather conditions and other factors likely to affect journey times. The ability to communicate with all other vehicles may not be a necessary precondition to achieve full autonomy but it would have enormous potential benefits. There are those within the industry who feel it would be essential in order to ensure absolute safety for autonomous vehicles, and they point to the fact that conventional cars are increasingly connected, and able to receive information externally. However, car-to-car communication technology is far more complex and is, as yet, undeveloped. To be useful and effective, the connectivity would need to be instant, with delays measured in microseconds, and fail-safe – requirements that are both onerous and expensive. There would need to be real changes, as mobile phone reception is poor in many places, and urban canyons of skyscrapers would make it particularly difficult to ensure 100 per cent reliable interconnectivity. There would need to be a massive investment in 5G and an ability to interact far more quickly than is currently possible. Paul Jennings reckons the car will become a platform for lots of apps (rather like a mobile phone today), 'many of which we cannot yet imagine'. However, he stressed that 'the interconnectivity agenda and the autonomy agenda are separate: they are not dependent on each other'.

All these revolutions are at an early stage and, of course, there is no guarantee that they will ever happen, either in the way predicted by the auto and tech companies or indeed at all. While difficulties over battery supply and the cost of vehicles are still issues, the electrification of the automobile fleet is almost certainly happening, not least because governments are increasingly regulating against polluting cars. However, there are still doubts about the capacity of the supply chain and the willingness of many car owners to transfer. There remains, too, a big question mark over whether the truck fleet can be run on batteries, but the trend towards greater, if not total, electrification is certainly unstoppable. Shared use of vehicles is increasing too, but it is likely to always be a limited market. The most important component of bringing about a driverless future is the development of the very sophisticated technology that will be required. The current state of that technology is examined in the next chapter.

Chapter 4

What can cars do now?

The advocates of autonomous technology are professional optimists. In an avalanche of speeches, press releases and blogs, they are forever assuring us that the brave new world of driverless cars is just around the corner. In announcements at motor shows and conferences, the tech and car companies, who are increasingly forming joint ventures to work together on autonomous car projects, make frequent references to the imminent prospect of full autonomy and suggest there will be a full fleet of cars by 2030 or so. The media laps this up, with journalists and 'experts' competing to be most bullish about the prospects. Yet even today's most advanced 'autonomous' vehicles are really the equivalent of the beautiful bangers that feature annually in the London to Brighton veteran car run when measured against the perfect driverless pods that will whisk people to work and then return to take the kids to school. The current state of the technology was well summed up in an article on the website Vox in early 2020:

> Despite extraordinary efforts from many of the leading names in tech and in automaking, fully autonomous cars are still out of reach except in special trial programs. You can buy a car that will automatically brake for you when it anticipates a collision, or one that helps keep you in your lane, or even a Tesla Model S whose Autopilot mostly handles highway driving.

The article goes on to say that every prediction from the auto and tech companies about the introduction of these vehicles has proved to be wrong.[31] The technology is currently hovering around Level 2 and Level 3, which is pretty much the same as when the first edition of this book was published in 2018, although the secrecy surrounding much of the development and the hype around information that is released to the public means it is difficult to get accurate information.

There is, therefore, a long way to go. This public relations effort from the auto manufacturers is necessary to justify the vast amounts they are spending on developing the technology but it presents a fantasy that is more *Blade Runner 2049* than *News at Ten*. R&D spending by the top twenty players in the automotive industry has soared in recent years: from $46 billion in 2015 to $70 billion in 2019. While it is not possible to disentangle precisely the proportion of that huge sum being spent on electric and autonomous car technology, according to the PriceWaterhouseCoopers report on connectivity cited in the previous chapter, 'the self-driving car will be the most valuable contribution to automakers' top and bottom lines in a generation'.[32] Therefore, it is highly likely that much of this money is being spent on the search for the Holy Grail of the self-driving car, and the actual sums mentioned by various companies back this up.

Predictions are constantly being made and updated – invariably pushed further into the future – and some of them will therefore be out of date or abandoned even before this book is published. In June 2017 the Venture-Beat website helpfully pulled together the promises of eleven top auto manufacturers in relation to their drive towards autonomy.[33] It is a litany of groundless optimism, baseless hype and, at times, pure bullshit.

Tesla's Elon Musk was by far the most bullish: in early 2017 he predicted that by the end of that year a Tesla would be able to drive from Los Angeles to New York City without a human touching the wheel. That does not mean, of course, that it would be driverless, and it would presumably all be on interstate highways rather than through urban areas. At the time of writing, in the summer of 2020, there is still no news of this plan, which has barely been mentioned since 2018.

General Motors, while eschewing a precise time line, is almost as gung-ho about its ambitions: 'we expect to be the first high-volume auto manufacturer to build fully autonomous vehicles in a mass-production assembly plant', its CEO, Marry Barra, said in December 2016. The company paid $581 million in 2016 to buy a self-driving technology start-up called Cruise Automation, and it has said its focus will be on ride sharing, suggesting that the company is predicting the decline of the current model of individual car ownership. GM's 'Super Cruise' system was released in 2018, but it is only considered as Level 2 on the automation scale. In June 2020, GM announced it was extending the introduction of its Super Cruise technology to twenty-two of its other vehicle models by 2023, but it will still only be possible to use it on the 200,000 miles of highways in the United States and Canada.

Several other car manufacturers have teamed up with tech companies too. Ford invested $1 billion in a robotics company called Argo AI, and, as mentioned in the previous chapter, it is skipping Level 3 automation. In February 2016, the company announced that it was aiming to have a 'fully autonomous vehicle' by 2021. In a statement issue in April 2020, Ford announced it was postponing the introduction of its 'self-driving services' until 2022, blaming the Covid-19 crisis. Meanwhile, Renault–Nissan is working with Microsoft to develop the company's autonomous car efforts and had plans to release ten different self-driving models by 2020. CEO Carlos Ghosn told the TechCrunch website that different levels of autonomy would be delivered over time:

> We know that autonomy is something of high interest for the consumers. This is the first brick – one-lane highway.

Then you're going to have multi-lane highway, and then
you're going to have urban driving. All of these steps are
going to come before 2020. ... 2020 for the autonomous
car in urban conditions, probably 2025 for the driverless
car.

Ghosn was famously sacked by Nissan in 2019 and
escaped house arrest by fleeing Japan hidden in a crate.
In June 2019, Renault–Nissan announced it was estab-
lishing a partnership with Waymo: 'to pursue development of self-driving systems for a range of vehicles that
will both carry passengers and haul packages'. However,
no timeframe was announced. In mid 2020, Groupe
Renault's website said that by 2022, '15 models will integrate autonomous driving capabilities'.

The other big Japanese companies Toyota and Honda
had both been aiming to have cars that 'drive themselves' in time for the Tokyo Olympics, which had been
due to take place in the summer of 2020: Honda also
paired with Waymo in order to achieve this, while Toyota
announced it was investing $1 billion in its autonomous
car project. Progress on these plans is unclear as no firm
announcements have been made by either company
since the games were postponed.

The South Korean firm Hyundai has also been investing heavily, with $1.7 billion allocated to its autonomous
car programme. In February 2017, the company's senior
research engineer, Byungyong You, said that they were

'targeting ... the highway in 2020 and urban driving in 2030',[34] which would effectively be Level 3 and Level 4, respectively. In 2019, Hyundai partnered with the Russian search company Yandex, which has been working on self-driving car technology. According to Yandex, the goal of the partnership is 'to create a self-driving platform that can be used by any car manufacturer or taxi fleet'. The 2020 deadline has not been met, however, and the company has made no further firm predictions.

Volvo is focusing on two different ends of the market: ride sharing and luxury cars. It has entered into a $300 million joint venture with Uber to develop autonomous cars. In an interview in July 2016, Volvo CEO Hakan Samuelsson said: 'It's our ambition to have a car that can drive fully autonomously on the highway by 2021.'[35] In May 2020, Volvo claimed it would have cars that could 'drive themselves on highways' in production by 2022. In June 2017, Daimler announced an agreement with Bosch to bring both Level 4 and Level 5 autonomous vehicles to urban environments 'by the beginning of the next decade'. Ola Källenius, Daimler's head of development, has said that he expects large-scale commercial production to take off between 2020 and 2025, but as of mid 2020 no firm date had been announced.

Not all manufacturers are as confident that the necessary technology is just around the corner, though. Fiat–Chrysler also linked up with Waymo in 2016 to test self-driving Chrysler Pacifica Hybrid minivans, but the experience led to scepticism from Fiat–Chrysler CEO Sergio Marchionne, who said he thought self-driving cars were further away than he once thought. He suggested that five years (2021) was a minimum, but interestingly he did not commit his own company to producing any.

In 2016, BMW announced a collaboration with Intel and Mobileye, a software company specializing in driver aids, to develop autonomous cars with the goal of getting 'highly and fully automated driving into series production by 2021'. Elmar Frickenstein, BMW's senior vice president for autonomous driving, said that the company should have Level 3 cars by that deadline but that it is possible they could even deliver cars with Level 4 or Level 5 capacity during that year. In the event, though, there is no sign of the company being able to meet this deadlne and it has gone quiet on its driverless car ambitions.

At successive motor shows, the manufacturers compete with one another to try to show who is most advanced in the race to autonomy. Their announcements are confusing, however, because while the technology is undoubtedly moving quickly, it is very difficult to disentangle the hype put out by manufacturers from the reality in relation to the current capability of their vehicles and what they are trying to achieve with this round of models. There is a disconnect between the achievements of test vehicles that cost hundreds of thousands – or even millions – of dollars to manufacture and the ability to turn those vehicles into workable models at prices that consumers will be able to afford.

Figure 4. Performance cars – in the driverless world?
(Photo by Falcon® Photography.)

In terms of electrification, the automotive companies are equally ambitious. BMW has said it will eventually have twenty-five models with electrified drives, half of which will be battery run, but it did not specify a timeframe. Daimler has promised electric versions of all its models by 2022.

In terms of autonomous vehicles, Audi is the most optimistic. In addition to its previously announced A8, said to have Level 3 capability as an option, the company announced it would be offering the Elaine, designed at Level 4, and, later, the Aicon, designed for Level 5. The latter boasted an interior that would not even require the passengers to face the road. Indeed, Audi's chairman, Rupert Stadler, said that Audi's aim was to use the technology to give customers a '25th hour', as well as an added living space, workspace or retreat. In other words, it would liberate the users of the car so that they would no longer in any sense be drivers and would therefore be able to use their journey in the same way as, say, train or plane passengers.

Motor shows like Frankfurt and Geneva, however, reveal the auto manufacturers' rather schizophrenic approach. While stressing that autonomy is the future, they are also introducing ever-higher-performing new models with the aim of attracting motorists who enjoy fast driving and rapid acceleration. The subtext seems to be that the boring self-driving pods will be for the hoi polloi while, at the top end, those able to afford high-performance cars will swan about in them at breathtaking speeds and in unprecedented luxury. The car makers seem to have failed to notice the fundamental contradiction between these conflicting visions of the future. They are riding two horses at once and trying to cater for two distinct and irreconcilable markets. Either they want people to treat cars – or rather mobility – as a utility provided by featureless shared pods, or they are sticking with the traditional market of 'driving as a pleasure', with high-end cars cruising along deserted roads. However, the benefits promised by the driverless future assume that no one will be driving their car themselves any more, which leaves no room – or rather road space – for the Mr Toads of the future.

The auto manufacturers are, therefore, hedging their bets. They still like the idea of selling top-of-the-range luxury cars that the owner is proud of and keen to drive. The self-driving world of entirely shared ownership and a utilitarian view of mobility that is supposed to free up car parking lots, and even possibly suburban roads, cannot be reconciled with the idea of retaining the conventional ownership model for drivers of high-end vehicles. Either we will have a world where the drivers of Lamborghinis and Ferraris can still, theoretically, enjoy the sort of motoring that is the preserve of car advertisements, or we will have one in which, as Elon Musk has suggested, 'owning a human-driven vehicle will be similar to owning a horse – rare and optional'.[36]

As a whole, the industry is predicting that a significant number of cars with some self-driving capacity will be on the roads by the early 2020s, with the first vehicles mostly being luxury cars or parts of commercial fleets. However, there is no consensus on when the true autonomy of Level 4 and Level 5 will be reached. Gill Pratt, CEO of the Toyota Research Institute, believes that 'none of us in the automobile or IT industries are close to achieving true Level 5 autonomy, we are not even close'.[37] His most optimistic prediction is that there will be a number of companies with Level 4 cars operating in specific areas within a decade.

There are two important considerations when looking at car company predictions. First, as the VentureBeat website points out:

> There are a lot of reasons for companies to be a little overly optimistic, such as generating national or company pride, earning media attention, boosting engineer recruiting efforts, and appealing to investors. There are few incentives, meanwhile, to be publicly pessimistic.[38]

The other consideration is that self-driving adoption time lines depend heavily on regulatory developments over the next few years. Autonomous vehicles require the right legal and technological frameworks. These will not only be complex to develop, particularly over liability issues, but they may also be hugely controversial. The nirvana of a world consisting solely of autonomous pods will require restrictions on other road users and personal car ownership, for example. While legislation has been enacted in many US states, and in several other countries, laws are still very much in an embryonic state and will require many future iterations.

For example, there are serious liability concerns when machines operate themselves in a potentially dangerous environment. Obviously, a car company doesn't have much incentive to mass produce a true self-driving car if there is nowhere it can be legally driven, or if the legal liability would be considered too risky. Uber already got itself thrown out of San Francisco when it attempted to test its Volvo cars there, and it ended up decamping to Arizona where the regulatory framework is looser. This led to a fatality.

Indeed, among potential fleet owners it is Uber that has been most eager to dispense with drivers. This is not surprising as Uber uses (not 'employs', since the whole Uber model is based on the legally questionable notion of drivers being quasi 'independent contractors') three million drivers, which represents a major proportion of the cost of providing a taxi service. In January 2015, the company claimed that entirely autonomous taxis would be available in 2018. At the time, as already mentioned, Uber's chief executive, Travis Kalanick, indicated in a tweet that he expected Uber's fleet to be entirely driverless by 2030. He added that the service would be so inexpensive and ubiquitous that car ownership would become obsolete.

In reality, the performance of the 'self-driving' taxis in Pittsburgh demonstrates just how far away the technology that would offer full-scale autonomy is. It was not only Uber's failure to provide jobs and its decision to charge passengers for the taxis that damaged its experience in Pittsburgh: the cars themselves proved to be far from autonomous. A leaked document revealed that in March 2017, when there were forty-three Uber 'driverless' vehicles in Pittsburgh, the average distance before an intervention from the 'driver' to take over was just 0.8 miles.

The cars drove a little over 20,000 miles that week, which means that there were 24,000 occasions when the 'driver' had to use the controls. The reasons for these interventions varied from the very serious – such as avoiding accidents – to minor matters such as unclear lane markings, the system overshooting a turn or bad weather. Accidental disengagements, when the driver took the controls by mistake, were excluded from the statistics.

Statistics from Waymo suggest a much lower rate of intervention but most of those miles have been travelled on highways, where there are far fewer hazards. There have been numerous incidents, and one of the early ones highlights both the limitations of the technology and the fact that 'driverless' is a term that is used very loosely. In 2016, an autonomous Google Lexus pulled into the path of a bus travelling at around 15 miles per hour, causing extensive damage to the car. The resulting statement by the company revealed the extent to which the 'driverless' cars are, well, driven. Google said that the test driver failed to intervene when the car made an error and pulled out in front of a bus: 'The Google AV [autonomous vehicle] test driver saw the bus approaching in the left side mirror but believed the bus would stop or slow to allow the Google AV to continue.' This was a bit of a giveaway. It showed that, far from being autonomous, many such vehicles still rely on human intervention to prevent accidents. It is the same reality that undermines those manufacturer safety statistics.

Waymo, which is way ahead of all other developers in the field, announced in December 2018 that a 'driverless' taxi service would start in Phoenix, Arizona, during the following year. The company had been testing its vehicles there since early 2017, on a commercial basis, but as with all these trials, the results and the lessons learned are couched in secrecy. Astonishingly, Waymo has insisted that people using the 'robotaxis' sign non-disclosure agreements that prevent them talking about their experience.

Again, 'driverless' does not really accurately describe the way these vehicles – which are anyway limited to a fifty-square-mile geofenced area of the suburbs around the city – operate. They are monitored remotely by a controller who is able to stop them, and there is a backup van with a spare driver in it to rescue any that have become stuck. Various reporters who have ridden in them speak of uncomfortable rides, sudden stops for obstacles such as pigeons, and an inability to deal with complex situations. They are very slow at the four-way stop signs that are prevalent in the United States and at T-junctions. One reporter who spoke to local people said that the most common reaction was 'I hate them', primarily because of the way they clog up local traffic.

Chapter 5

Bumps in the road

So what is wrong with the various utopias presented by the auto and tech companies? Let me make something clear from the outset. I am not going to bother with the oft-posed dilemma during debates on autonomous cars about how to programme the car to choose between, say, killing an elderly pedestrian or piling into a bunch of schoolchildren when faced with a certain accident. For a start, it is a purely theoretical situation that will occur very rarely. Humans might, in any case, make the same wrong decision in the split second available to them, which makes imposing the requirement for a driverless car to be able to negotiate this conundrum superfluous. More importantly, if that were the only issue facing autonomous cars, they would be ready to be introduced widely since it is such a rare occurrence. As we have seen, though, there are no end of far bigger and more relevant issues that need to be considered before we get into that rather too frequently posited conundrum, which is best left to late-night drunken discussions.

First, there is the haziness of the vision, especially given that any future method of transportation will have such an enormous impact on the type of society we – or, more importantly, our children and grandchildren – will inhabit. Transport is such an intrinsic part of all our lives that introducing a fundamental change like the development of autonomous cars requires very careful consideration by governments and civil society. The technology itself should not be allowed to drive change. The various visions of the future range from a completely driverless world in which individual car ownership is unknown or even banned to various hybrid versions in which driverless and driven cars will share road space or, perhaps, will be separated, with the former getting priority.

Even more hazy is how we get there. The key point is that the development of autonomous cars will not be a gradual process through which existing vehicles will have more driver aids added that will take them through the various stages of autonomy until they reach Level 5. Instead, the vehicles will need to be completely different from existing models, as Google has now recognized. Therefore it will need a step change, not a gradual incremental one. No one has any idea, so far, of how this could be managed, making the talk of an imminent driverless future a nonsense.

Before we can even contemplate a 'road map' that would lead to the widespread adoption of autonomous cars, a series of major hurdles must be overcome.

Let us look first at **safety**, since this is the overriding consideration set out by the tech companies and auto manufacturers in support of the introduction of these vehicles. Despite some claims that Level 3 would have a positive impact on safety, it is already pretty clear that it is not until Level 4 and probably Level 5 is reached that the promised benefits will be delivered. And even then, the technology itself has to be extremely safe and not hackable.

The problem with the testing regime of the vehicles is that the trials are carried out in unrealistic conditions. Tesla responded to the accident that killed Joshua Brown (mentioned in chapter 3) by arguing that statistics show there is a road death for every 100 million miles driven whereas 130 million had already been driven in Autopilot mode up to the point of the Florida accident. One casualty, therefore, was statistically positive, representing a reduction compared with human drivers. That is totally misleading. The Autopilot mode is only suitable for highway driving, which is far safer than average and consequently not a precise indicator. Moreover, those statistics would not reveal when an accident took place in the immediate period after a driver had disengaged autopilot, precisely because of fears of an imminent collision. Nor, crucially, would they record the number of times drivers had disengaged the Autopilot to prevent an accident. The RAND Corporation calculated that to show that self-driving cars were as safe as human drivers would require 275 million fatality-free miles, and even that might be an underestimate.

Contrary to the arguments set out by the proponents of autonomous cars, human drivers are actually pretty good. We dodge most obstacles most of the time and anticipate danger well, especially those of us with a few thousand miles under our belts. Rob Dingess of Mercer Strategic Alliance, a lobbying firm specializing in automobile technology, who has a quarter of a century of experience studying transport technology, summed it up well: 'While human factors largely contribute to the failures referred to as crashes, human drivers avoid such incidents in relatively complex environments on a miraculous scale.'[39] It should be noted, too, that it will be the same error-prone human race that will provide the devices and the software that the autonomous cars will need. Dingess also observed, with irony, that the manufacturers had become very good at 'developing self-driving systems that operate safely 90 per cent of the time, but consumers are not happy with a car that only crashes 10 per cent of the time'.

In evidence to the March 2016 US Congressional hearings on automated cars, Missy Cummings – whose job title at Duke University is 'Director, Humans and Autonomy Laboratory' and who is a self-proclaimed supporter of the technology – expressed strong doubts about the way that safety testing was being carried out and overseen by the regulators. She pointed out that whereas the Federal Aviation Administration has set out a clear certification procedure for aircraft software, and would not allow commercial planes to make automatic landings without having verified that software, by contrast, she said, 'any certification of self-driving cars will not be possible until manufacturers provide greater transparency and disclose how they are testing their cars'. She added that 'they should make such data publicly available for expert validation'.[40] In fact, quite the opposite is happening. Manufacturers are not sharing information or divulging it, through fear of passing information to their competitors, and they are not providing it to regulators.

The National Transportation Safety Board (NTSB) investigation into the accident that killed Joshua Brown revealed that the Tesla Model S uses the company's own proprietary system to record a vehicle's speed and other data. This means that external agencies cannot access this information using the usual tools available commercially for gathering data from most other cars. For that reason, the NTSB said it 'had to rely on Tesla to provide the data in engineering units using proprietary manufacturer software'.[41] That could mean that potential faults in the system might not be revealed, and no independent verification of the safety of the system is possible.

It was this secrecy and lack of information that led to Cummings expressing concerns that autonomous vehicles, albeit still with test drivers at the wheel, were being let loose in many US cities without sufficient consideration for the safety implications. There was, she suggested, a lack of accountability that could put people at risk. As she said, 'inability of self-driving cars to follow a traffic policeman's gestures, especially on a rainy day in a poncho, means that self-driving cars should not really be operating near elementary schools at this time'.

Cummings also addressed the issue of **hacking** at the same hearing. She gave the following example:

> It is relatively easy to spoof the GPS of self-driving vehicles, which involves hacking into their systems and guiding them off course. Without proper security systems in place, it is feasible that people could commandeer self-driving vehicles to do their bidding, which could be malicious or simply just for the thrill.

This issue has been of great concern to developers of autonomous cars, and an article in the *MIT Technology Review* outlined details of various forms of hacking that could disrupt autonomous vehicle use:

> [Autonomous] vehicles will have to anticipate and defend against a full spectrum of malicious attackers wielding both traditional cyberattacks and a new generation of attacks based on so-called adversarial machine learning.[42]

The author points out that one possible motive, apart from terrorism, for cyber attacks on autonomous cars would be anger over the widespread loss of jobs that would result from their introduction (an issue that we discuss in the next chapter). He warns that before self-driving taxis can become a reality, the vehicles' architects will need to consider everything from the vast array of automation in driverless cars that can be remotely hijacked to the possibility that passengers themselves could use their physical access to sabotage an unmanned vehicle being used as a taxi. The vehicles would have to include a port (called an 'OBD2') that enables mechanics to diagnose malfunctions in the car's systems, and that would allow hackers to alter a vehicle's control mechanisms. In an experiment in 2014, testers remotely hacked into a Jeep Cherokee, disabling its brakes – a breach that was replicated in a trial the following year. One of the hackers was Charlie Miller, who worked for Uber at the time but left in order to be able to share the information he gained in these experiments, and he warned that before self-driving taxis could be introduced, the manufacturers would need to consider everything from the various parts of driverless car automation that could be remotely hijacked to the risks posed by malevolent passengers:

> Autonomous vehicles are at the apex of all the terrible things that can go wrong. Cars are already insecure, and you're adding a bunch of sensors and computers that are controlling them. ... If a bad guy gets control of that, it's going to be even worse.[43]

Another significant issue is the possibility of people playing **pranks**. Missy Cummings pointed out that a $60 laser device could be used to trick autonomous cars into sensing nonexistent objects. The GPS systems could be disabled too, she said:

> It is not uncommon in many parts of the country for people to drive with GPS jammers in the back of their trunks to make sure no one knows where they are, which could be very disruptive to the system.

Researchers at the University of Washington managed to confuse autonomous cars into misidentifying road signs by using simple stickers they overlaid on them, fooling the car's image-detecting algorithms into thinking they were seeing, for example, a speed limit sign instead of a stop sign. There will clearly have to be legislation preventing people from playing such pranks, but that does not allay the security concerns raised by such experiments.

Indeed, another form of interference, but one that could also be used by **criminals**, seems to pose almost insuperable problems for the idea of a fully autonomous fleet. By definition, an autonomous car will have to stop to avoid hitting a pedestrian. The technology does allow for the distance between the pedestrian and the vehicle to be related to the speed at which it is travelling. However, autonomous cars will have to be programmed not to run people over. Therefore, they will have to stop when there is a person in the road. (Incidentally, trespassers on the line cause frequent delays to train services for the same reason: railway companies do not want to be responsible for killing them when they are on the tracks.)

This raises fundamental issues. At one end, there are the kids who might run in front of a car just as a joke, in the same way I used to put conkers in the street so that cars would splatter them. My efforts did not prevent the cars continuing on their journeys, though. Then there is what I dub 'the Holborn problem'. Every weekday evening there are so many people around Holborn Tube station in central London that they spill into the very busy Kingsway and High Holborn streets. It is nightly chaos. If these people discovered, however, that the cars could not move until they got out of the way, there would be permanent gridlock. The people would take over the streets. The only solution therefore would be far stricter restrictions on pedestrians, something that in an urban context would be highly damaging, turning all streets into priority zones for vehicles.

At the most extreme end, there is the problem of what the gun lobby likes to call 'bad people'. If autonomous cars have to stop because there are people in the road, then travelling through unsafe areas becomes impossible. It would be far too easy for the vehicle to be forced to stop, with its occupants then being attacked.

To go further with this logic, no one concerned with their security would be willing to risk travelling in a vehicle over which they have no control. This rules out the notion that government ministers and other VIPs will ever travel in autonomous cars.

The Highway Code for autonomous cars would have to be very different from the current one for conventional vehicles. Consider a road where there is a single lane because of parked cars on both sides, as happens often in older European cities. A human driver is able to assess the potential danger of a child or other pedestrian stepping out into the road and drive at a reasonable speed. An autonomous car is likely to have to go slower because of the impossibility of assessing what is going on behind the cars and distinguishing between, say, pedestrians walking on the pavement and those who might be about to move towards the road. (In my own experience, I remember as a teenager driving too fast in such a circumstance and killing a cat that crossed in front of me. My hands had twitched to steer away when it ran in front of me before I instantly recognized that I would crash if I tried to avoid the poor animal. Would an autonomous car really ever be able to make such decisions, identifying the difference between a cat and a child, perhaps one wearing a dark coat, instantly?) And what will happen when two cars face each other in such a situation? How will the autonomous cars negotiate who should reverse to let the other one through?

In a more banal but similar situation, what will happen at a T-junction, say, where the main road is extremely busy and cars are therefore having difficulty coming out of a side road. As humans, we creep forward and eventually a kindly fellow motorist will flash their lights to let us out once our eyes have met. Autonomous cars will not be able to do that because they will be programmed not to creep towards danger since that will inevitably cause crashes.

Such issues are not trivial. There are some signs of a growing awareness, in private at least, among the auto manufacturers and technology firms of the scale of these difficulties. Dingess has written about returning from the 2017 conference of the Transportation Research Board Automated Vehicle Symposium in San Francisco having found that it was beginning to dawn on many indus-try enthusiasts for the technology that the mountains they had to scale were higher and steeper than they had previously thought. He noted that 'whereas a few years ago, the industry operated under cautious opti-mism, today the discussion is far more pessimistic'.[44] The reason for this pessimism is all too clear: the technical requirements for fully autonomous vehicles are likely to be harder to achieve than those needed to send a rocket to land on Mars.

This is because to even begin reaching the nirvana promised by the tech and auto industries, they would need to produce a car that could work everywhere, all the time, dealing with all eventualities. Not 98 per cent or 99 per cent of the time – let alone 90 per cent – but 100 per cent of the time. At the moment, the sensors find it difficult to cope with fog and, particularly, heavy rain. Snow is a problem too, both as it is falling and, possibly more seriously, when it has settled, when it changes the landscape and, crucially, masks the road markings that are a vital part of programming navigation systems. Without good road markings, the current generation of cars will need white sticks. As a major US driverless car website put it, 'nobody has even started to address the problem of how to navigate in snow-covered areas, which may in turn have implications on the mapping approach'.[45] Road markings, incidentally, are one of those external costs that will be borne not by the car manufacturers but rather by third parties such as local authorities. If, as considered earlier, cars are expected to communicate with each other and possibly with features in the road, there may well be considerable extra costs incurred by cash-strapped councils to accommodate autonomous vehicles. There is also the problem of GPS canyons and blind spots, created by rows of skyscrapers or other natural or human-made features. Remember – and it is worth reiterating this time and time again – that the whole driverless concept presented to us by the ever-enthusiastic tech companies only works if cars can go everywhere from desert trails and paths through fields to tunnels and streets in business districts sand-wiched between megablocks.

Figure 5. Cyclists, pedestrians and snow: computer says 'no'?
(Photo by Jim Henderson.)

There are a plethora of **other technical challenges** that would also need to be overcome. Take cyclists. Carlos Ghosn, Nissan's former CEO, is no fan of people on bikes. Indeed, he hates them. In a speech in January 2017 to introduce a prototype 'driverless car', he told CBNC that the arrival of the technology could be delayed by cyclists who, he said, 'don't respect any rules usually'. Ghosn reckoned this was a particularly difficult nut to crack for progress towards autonomy:

> One of the biggest problems is people with bicycles. The car is confused by [cyclists] because from time-to-time they behave like pedestrians and from time-to-time they behave like cars.[46]

The implication was that he would like to see cyclists banned from roads where autonomous cars are travelling. Just as in my Holborn problem above, there is a strong line of thought among promoters of autonomous technology that other road users will simply have to make way for their vehicles.

When Nissan took a group of journalists for a ride in an autonomous car in London a couple of months later, the reason for Ghosn's antipathy became all too clear. The passengers noticed that the car had passed too close to a cyclist it was overtaking and by chance a French journalist happened to film the incident. This was highly embarrassing for the company as the event was supposed to be a demonstration of how well the 'Nissan Intelligent Mobility' autonomous car could integrate with existing traffic on a dual carriageway in east London. The car – a plug-in electric Nissan Leaf – was guided by five radars, four lasers and a dozen cameras, but despite this panoply of equipment it did not move into the adjoining lane, which was empty, as it should have done while overtaking the hi-vis-jacketed cyclist. As the BBC transport journalist who was in the car noted, this was in breach of rule 163 of the Highway Code, which states that motorists should give cyclists, as well as pedestrians and horse riders, as much space as they would give a motor vehicle while overtaking.

This is not the only traffic rule that these quasi-autonomous cars break. It is clear that they are not programmed to obey speed limits. This is another quandary for their developers. If they were speed limited, other vehicles would pass them on motorways where speeds are regularly greater than the 70 miles per hour limit. They would, in fact, cause a hazard, as other vehicles might weave in and out in front of them, knowing that they could not accelerate higher than the speed limit. However, this implies that the software would allow the car to break the law – as was the case, in fact, with the car being not-driven by Joshua Brown, the Tesla enthusiast, when he met his tragic end. In other words, autonomous cars will be damned if they do break the law and damned if they don't.

This highlights a key obstacle to the widespread introduction of autonomous cars. The transfer from human to autonomous control cannot happen overnight and this means that there will be a long period during which driverless and driven vehicles will share the roads, posing many challenges for those driverless vehicles. Even the most optimistic of driverless car advocates accept that this period will last many years, if not decades, and some accept that it is likely to be permanent. The notion that everyone – people living in rural areas or off the beaten track, or those with special needs such as baby seats in the back, or those requiring particularly big or particularly small vehicles, and so on – will eventually choose to have a driverless shared car is unrealistic. As well as the example cited above about VIP travel, there are numerous other problems to get round. Ambulances, police cars and fire engines are unlikely to be transferred to autonomous mode. Delivery drivers are likely to remain too – *pace* Amazon's unworkable fantasy of drones knocking on people's doors in ninth-storey flats to drop off their parcels – because humans will still be needed to carry the goods to their recipients and make decisions about what to do when no one is home. Deliveries to shops using autonomous vehicles may be more feasible, but there are still numerous potential difficulties, such as coping with complex areas of private-access roads guarded by automatic gates. My favourite example is: who will allow a driverless car to take them into a safari park?

The platooning of lorries – an experiment that was scheduled to be tested on British motorways in 2018 but after several postponements now has, as of mid 2020, no scheduled date – also potentially poses risks to other road users. The idea is that by keeping the lorries close together they will use less fuel, and though this was of course not stated by ministers when announcing the £8 million grant for the project, it will also reduce the number of drivers (although what would happen at either end of the motorway is unclear). Other motorists may well be tempted to weave between the lorries, particularly if they need to turn off the motorway and find their way blocked by half a dozen trucks. Edmund King, the president of the AA, is not impressed with the plan:

We all want to promote fuel efficiency and reduced congestion but we are not yet convinced that lorry platooning on UK motorways is the way to go about it. We have
some of the busiest motorways in Europe with many
exits and entries. Platooning may work on the miles of
deserted freeways in Arizona or Nevada but this is not
America. A platoon of just three HGVs can obscure road
signs from drivers in the outside lanes and potentially
make access to entries or exits difficult for other drivers.
Even a three-truck platoon is longer than half a Premier
League football pitch.[47]

Here there is another dilemma for manufacturers
and, in particular, regulators, who face a daunting task
in working out how to create a safe environment for the
potential widespread introduction of autonomous cars:
in order to stop other drivers weaving in and out of lines
of autonomous cars, the vehicles' software will have
to be designed to make them assertive, even possibly
aggressive. In the most dystopian view, manufacturers
will compete against each other to make their cars the
most able to deal with traffic, in which case the regulators would clearly have to step in. On the other hand, it
may simply be too dangerous to allow autonomous cars
to behave and react in that way. All of which reinforces
the point made by Cummings about the need for the
regulators to have full access to the computer systems
in order to assess the risks contained within them and,

particularly, to see what parameters have been set for the software designers.

Autonomous cars will make use of technology that enables machine learning. In other words, they will gradually be trained to recognize certain objects and react appropriately. Early tests of Google cars were hampered by the cars reacting to inanimate objects such as plastic bags blown by the wind. A 2014 article in *Tech Times* highlighted this problem:

> The car has yet to be tested for driving in snow and during heavy rains, along with traversing wide parking lots and multi-floor garages. The self-driving car sees people as moving pixels, which means that it will not recognize a police officer waving frantically for all cars to stop. The sensors of the car will also not be able to tell if an object on the road is a rock or a plastic bag, so it will be driving around both of them.[48]

This early overresponsiveness to inanimate objects led to several rear-end shunts in the initial phases of Google testing, as human drivers failed to understand why the car in front was making an emergency stop. Despite this, Chris Urmson (who was the head of the testing programme at the time) was quoted in the article as being confident that the vehicles would be publicly available by 2019, when his son reached the age of 16: the minimum driving age in California. (They were not, of course.) Although Waymo (as this division of Google is now known) claims this problem has largely been alleviated thanks to machine learning, no independent verification of this is available. As with other developers of the technology, the real-time capability of their machines is a closely guarded secret. And as we have seen, if inanimate objects are a problem, animate ones present a much greater challenge.

Figure 6. Unplanned roadworks: bad enough for an autonomous vehicle, worse still if the fencing were not there. (Photo by Editor5807.)

The other issue highlighted during the early days of Google testing remain problematic. Driverless cars are all heavily dependent on mapping, and consequently they can only be used where their software has been loaded with details of the local road conditions. Moreover, unexpected changes – such as a new traffic light, an altered road layout or unplanned roadworks – require the test driver to take back control of the vehicle. This is a source not only of potential inconvenience, with autonomous cars stopping at unexpected hazards, but of safety, as they fail to recognize them.

This brief round-up of the known unknowns is sufficient to show that the technology faces a myriad of difficulties before its widespread introduction, and that is even before, to continue quoting Donald Rumsfeld, we have come to the unknown unknowns. In that respect it is fascinating to contrast the development of driverless cars with railway technology. In the mid 1990s, when the railways were being privatized, Railtrack, the predecessor of Network Rail that had responsibility for the infrastructure of the system, confidently asserted that trains on the West Coast Main Line, the busiest mainline in the network (serving Birmingham, Manchester, Liverpool and Glasgow), would be controlled by in-cab signalling rather than external lights. This, the company promised, would deliver a huge increase in capacity as trains would be able to operate closer together and at higher speeds. Under the contract it signed with Virgin, most of the improvements were to be delivered by 2002 when the line would be fully fitted with in-cab signalling. But none of this ever happened. The technology has never been introduced on complex routes like the West Coast Main Line, with its myriad junctions and wide variety of different trains (suburban, long distance, freight, etc.).

The scheme was soon abandoned and Virgin received hundreds of millions of pounds in compensation.

There are clear parallels here. The new technology (or different versions of it) is used in many simple railway systems, such as underground lines, but the challenge of trying to introduce it in more complex operating conditions has so far proved insuperable, despite the encouragement of the European Union (it is called the European Railway Traffic Management System). Similarly, there will be niche uses for driverless cars – like the bus that trundled around the Olympic Park for a few months in 2018 at 5 miles per hour – but the step to widespread adoption is even greater than the challenge that Railtrack faced, and failed to meet.

Privacy and control of personal data are also going to be major points of contention. These cars carry cameras that look both inside and outside the car, and they will transmit both these images and telemetry data in real time, including data about where you are going and your driving habits. The issues of who has access to this data, whether it is secure and whether it can be used for other commercial or government purposes have yet to be addressed.

Let us close this chapter by looking at three final considerations. Autonomous cars are going to require a very clear **regulatory framework**. In the United States, much of this has been delivered at the state level, causing widespread confusion and difficulties for the prototypes being tested on the roads. It is likely that in Europe, legislation will be developed at the pan-European level in order to enable vehicles to drive across frontiers without delay. Now that Britian has left the European Union, this may prove an even greater obstacle to the widespread introduction of these vehicles. A clear set of safety standards will be required. As Missy Cummings has pointed out, this will require much more openness on the part of manufacturers about the capability of their machines, and tests of whether the cars meet the standards will be needed.

As Tesla learned in the aftermath of Joshua Brown's accident, the **legal issues** are mind-boggling. Who is going to be allowed to 'drive' these vehicles and where? Codes of practice for 'driverless' vehicles on public roads have existed since 2015 in the United Kingdom, and they address a number of issues relevant to the testing of these vehicles – from vehicle and test driver requirements, to insurance, data protection and cyber-security issues – but primary legislation will still be required when privately owned vehicles hit the road. There have even been suggestions that special permission will have to be granted to people who want to keep driving themselves.

The other challenge relates to **liability**. Who will be at fault if a driverless car piles into another vehicle? Will it be the manufacturer, the software provider, the owner, the council that paints the road, or the regulator?

Google and Tesla are pushing for a legal framework for their vehicles to ensure that once the technology is developed, it will be possible to sell them straight away. However, the regulators may be far more cautious than the tech companies want because of public opinion and the reluctance of the companies to provide information. Volvo has attempted to pre-empt the situation by accepting liability for any collisions involving its autonomous vehicles. This is an easy promise to make when there are no cars on the road, but it might be far more difficult if there were a series of incidents that could bankrupt the company. Hacking could also cause accidents, adding complexity to the question of liability. Will autonomous cars require regular 'patches' to make up for security or other flaws? And if so, whose fault will it be if the patch is not installed?

Even a small-scale trial involving three autonomous vehicles (with a 'driver' aboard) on fifteen miles of road in an older citizens' residential village in San Jose, California was nearly killed off because of insurance concerns. According to an article in the *New York Times*,[49] the insurance company Munich Re demanded twice the normal premium to cover the cars for payments of up to $5 million. A compromise was eventually reached, with the driverless car company, Voyage, agreeing to provide data to the insurers as well as a higher fee. Interestingly, the article says that the village was a good testing ground for Voyage: '[because] it is private property, the company does not have to share ride information with regulators and it can try new ideas without as much red tape'.

Just because a technology exists – and autonomous cars don't at the moment – does not mean it will be used. As an example, look at helicopters. It is perhaps surprising that they have not been more widely adopted by the rich for transportation into cities. Robert Maxwell, for whom I worked briefly in the 1980s, was in the habit of flying in one between his Oxfordshire home and the *Daily Mirror* office in Holborn. However, he was very much an exception in having somehow obtained a dispensation from the aviation authorities. For the most part, the disbenefits of noise and risk to the general public caused by helicopters far outweigh the advantages to a few high-net-worth individuals, and city authorities across the world have consequently effectively killed off the idea of helicopter commuting.

Another example is the Kindle. Remember that this technology was supposed to eliminate books, but despite the proliferation of devices (including the simple smartphone) on which it is possible to read, after a bit of a blip when e-readers first became available, book sales started to rise again. Kindles are rarely even seen on trains or planes, where they have a distinct advantage, let alone in people's living rooms. They are a minority market, and it is quite possible that despite all the gung-ho promises from the auto and tech manufacturers, autonomous cars will never, as it were, take off. And this is not necessarily for any of the various reasons stated above but simply because people will not take to the idea or because many will believe that the disadvantages outweigh the benefits. Edmund King makes a point that is surprising to many urbanites: people still like driving. As he says, 'In a Populus survey of 26,000 AA members, 63 per cent said they still enjoy driving and 69 per cent said they are not ready to take their hands off the wheel.'[50] Attitudes, of course, could shift, but there will be considerable consumer resistance to overcome. Even anecdotally, more people mention a reluctance to use this technology to me than get excited about its potential. A survey in April 2020 for the Smart Transport website found that very few UK drivers would trust driverless technology. The survey of 2,000 drivers commissioned by InsuretheGap.com found that only 8 per cent of respondents would feel safe travelling in a fully self-driving vehicle and that a further half would want to have the option of taking over the controls.[51]

One rarely mentioned further obstacle is cost. For the moment, the vision of an urban landscape dominated by autonomous cars remains a techie's dream, and the proposed utopia is being funded by the vast super-profits made by tech monopolies and by auto manufacturers' huge R&D budgets created out of a fear of being left behind. Google's cars cost at least $150,000 each, probably much more, including more than $70,000 just for the spinning laser rangefinder – the LIDAR – on the roof of each one. With economies of scale, costs will fall of course, but there is no guarantee that the pods, or whatever emerges, will be affordable, even to fleet buyers who will have to make a return on capital.

For the sake of balance, though, I should address head on the point made by Elon Musk and others that even if autonomous cars are not 100 per cent safe, as long as they offer significant enhanced safety, thousands of lives will be saved and therefore they should be introduced even if the technology is not perfect. As we have seen, it is unclear whether autonomous cars will actually be much safer in real-life conditions. It is only when they can operate in snow, or on huge empty parking lots, or on dirt roads, or where there are roadworks, or when an ambulance needs them to get out of the way, or in a dust storm, or when there is a fallen tree blocking the road, or when a troop of monkeys is playing on a road, or in floods – or in a myriad of other circumstances that humans can deal with quite well – that their relative safety can be measured. The partially driverless car – such as one that operates at Level 3, which Ford has abandoned – may well be less safe than the current situation. Having a mix of driverless and conventional vehicles may not be optimal for safety either. The claims made by supporters of the autonomous technology have not, therefore, been tested. Of course it is true that most accidents are currently caused by human error, but there is no guarantee that the new technology will eliminate those accidents, even though they may appear in a different form. Indeed, doubts are creeping in about the safety claims of the main proponents of the technology. Research into the causes of 5,000 crashes by the US Insurance Institute of Highway Safety published in June 2020 found that only a third would have been prevented by driverless technology. The researchers found that only accidents caused by direct driver error, or by their 'incapacitation', would have been prevented had there been no driver. Other types of accidents, such as cyclists pulling in front of vehicles, would not be stopped.

Moreover, to achieve better performance than the current safety performance of conventional vehicles, designers of the software for driverless cars would have to set the parameters for safety much higher, meaning that it would have to be stipulated that the cars would operate at lower speeds than conventional vehicles.

It can and will be argued that all these problems are solvable, of course; that the technology will somehow overcome even such conundrums as how to deal with 'bad people' standing in front of vehicles. However, it may also be the case that the obstacles are simply insuperable, or that they would cost so much to overcome as to make it not worthwhile to do so. The tech and auto manufacturers refuse to address this issue, and their secrecy about their progress – or lack thereof – only adds to doubts. Why should we believe their promises when what we have so far seen is so far away from the new world they are presenting to us? Elon Musk can talk all he wants about his 'Autopilot' system, but how many more Joshua Browns will it take before the whole house of cards collapses. The fable of the emperor's new clothes comes to mind.

Perception is important, and ultimately that is what will determine the fate of driverless cars in the unlikely event that all the other problems can be resolved. As Simson Garfinkel wrote in the *MIT Technology Review*:

> It will take only a few accidents to stop the deployment of driverless vehicles. This probably won't hamper advanced autopilot systems, but it's likely to be a considerable deterrent for the deployment of vehicles that are fully autonomous.[52]

The perception of risk is crucial. People will not put themselves at risk if they are at the mercy of malfunctioning machines. Drivers feel that they are in control, and that for the most part their skills enable them to avoid danger. They may be wrong, but it remains a key barrier for the supporters of autonomous cars to cross, even if all the other challenges cited in this chapter can be overcome.

For a bit of fun let us end this analysis with the predictions made in an April 2017 *Times* article[53] based on a blog by US-based Benedict Evans. The article was written by the paper's technology correspondent, Mark Bridge, who assumed that autonomous cars would become commonplace by 2030. Bridge lists ten advantages under the headline 'How driverless cars will fuel a better world'.

- **No more congestion.** Bridge quotes a Department for Transport brief that suggests that if penetration of driverless cars reached 25 per cent, congestion delays on urban roads would fall by 12.4 per cent. **However...** There may well be more traffic generated by driverless cars, rather than less. The next chapter deals with this issue in more detail.

- **No accidents.** This would save £34 billion per year, the article claims. **However...** There is no guarantee that this technology will be safe, or that it will ever be introduced in the way that has been presented.

- **Cab fares would be only £1,** because there would be no drivers to pay. **However...** There is no certainty that driverless vehicles will ever be able to operate in complex urban environments, with skyscraper canyons and countless human beings flooding onto the roads. Even if they did, costs would not necessarily be driven down, particularly if Uber or one of its rivals creates a local monopoly.

- **Green streets.** There will be no need for parked cars since shared cars will not be owned by residents. **However...** As we have seen, this is a very unlikely outcome of this technology.
- **The end of public transport.** Cheap on-demand transport would displace public transport and some bus routes could become obsolete. Eventually, according to Evans, 'if most people can afford on-demand car services, there's a risk governments [would] remove public transport networks'. **However...** This would not be a cause for celebration and would in fact contradict the idea that congestion would be reduced since many more people would be using individual vehicles, clogging up the roads.
- **Different geographies** would be created, since areas with poor transport links might become far more desirable. **However...** This is quite possible but wholly unpredictable.
- **The elderly will become more active** as they will be able to get around more. **However...** The fittest and healthiest older people are those who use public transport and walk. Autonomous technology will merely encourage them to continue using cars.
- **Property will become cheaper** because urban developments will no longer need parking facilities, which can push costs up by 20 per cent. **However...** Apart from the fact that this assumption is, again, based on the introduction of shared pods, the suggestion that

property developers would no longer need to provide for parking is a planning decision, not a transport one, and is not dependent on the technology.

- **More boozing** but...
- **... fewer snacks/cigarettes,** since these are frequently purchased at petrol stations. **However...** Do me a favour! Neither of these is a serious point.

Chapter 6

Who will drive the Queen?

It is fiercely difficult to predict the impact of technology. The promise of a congestion-free world of driverless pods is seductive but, as we have seen, it can easily be shown to rely on a specific series of developments and societal changes that are unlikely to come about. Before society begins to stake transport policy's future on the dream of autonomous cars, and before investors bet their life savings on Elon Musk's technological promises, politicians and regulators need to take a deep breath.

Paul Jennings of Warwick University – who is a considered and serious proponent of autonomous cars – told me in an interview why he is so keen on helping the development of the technology. He cited three main reasons. First, the safety argument, using the statistic that 90 per cent of road deaths are caused by human error. Then, in his words, 'with a more intelligent control of the vehicle, we can potentially save energy, we can potentially improve air quality, we could improve congestion, though there is a debate about that, and [additionally] people like their independence: having these vehicles will enable some people to be mobile who do not have licences'. He added that it might also free up time, 'as on average motorists spend six working weeks, 230 hours, per year on the road'. He does accept that these advantages will only come 'if society gets it right'. And therein lies the rub. I asked him who was pushing the societal interest, and after a long pause he said that 'it is coming in some of the research'.[54] Moreover, all his predictions were couched in cautious language about the length of time major developments might take to be realized and how long it might take for the technology to be fine-tuned.

One of the potentially far-reaching advantages suggested by Jennings was the relief of congestion, although he accepted there was no guarantee that this would come about as a result of autonomy. After safety, the reduction of congestion is the second most commonly cited advantage of the technology. However, even if the promise of an entirely autonomous future is achieved, the prediction of reduced traffic cannot be sustained. One reason given for the claimed reduction in congestion is the fact that autonomy will allow cars to drive just a few feet apart, even at 80 miles per hour, which will allow road space to be used far more efficiently. Setting aside the point that speeds above 30 or 60 miles per hour are not allowed in the urban areas where there is most congestion, the laws of physics still pertain. A lengthy separation, as set out in detail by the Highway Code, would still be required in case a vehicle in front came to an immediate halt; otherwise there would be an unimaginable pile-up. And that possibility – of a car suddenly stopping – cannot be eliminated when anything, such as debris blown by the wind, or a dog, or a rock, or a person, might end up on the highway, and of course a vehicle might simply run out of juice or break down.

Again, a comparison with the railways is instructive here. Modern signalling techniques allow more trains to use the tracks because they use a 'moving block' system. In other words, rather than having fixed sections of track (blocks) controlled by signals in which only one train can be at any time, in a moving block system, trains are controlled by signals sent into the cab, which means they can be closer to the train in front at slow speeds. However, they must always be able to stop in time to avoid a collision should the train in front stop instantly, because of a derailment, say. The capacity of the line is therefore constrained by the speed limit, and this would be the same for vehicles. No rational person would agree to travel in a vehicle at 80 miles per hour if it was only a few feet away from the vehicle in front of it.

Another argument used to support the idea that congestion will be reduced is that since most cars spend most of the day parked, and are used only 5 per cent of the time, perhaps as many as 90 per cent of vehicles could be dispensed with altogether. This ignores the fact that a high proportion of vehicles are in use at the same time (peak times): there will remain very little demand for transport at 3 a.m.! While it is true that there might be a small reduction in demand in some circumstances, there may also be an increase in the total number of journeys, because driverless cars might operate without any occupants as they are sent to park or pick people up (known in the trade as 'zombie miles'), and because of the additional number of people who are able to use cars, such as those without licences or people with poor eyesight. Indeed, at peak times one can easily envisage a far greater tendency towards gridlock than exists today, as thousands of empty vehicles vie for space with occupied ones. As with so many of the predictions of the benefits of autonomy, the notion of reduced congestion is predicated on the idea that we will abandon individual car ownership.

Yet, as we have seen, full autonomy, let alone universal shared use, is an impossibility. As the *reductio ad absurdum*, just imagine the Queen travelling in a driverless pod that might have been used by a vomiting drunk the night before! As a House of Lords committee concluded:

> The theoretical potential of CAV to reduce traffic congestion varies depending on the level of vehicle autonomy and the penetration rate. While we cannot say with any certainty what the impact on congestion will be, it is possible to imagine a situation of total gridlock as Connected and Autonomous Vehicles crawl around city centres.[55]

The impact of autonomous vehicles on employment is another issue that is little considered by the proponents of the new technology. There are around 600,000 HGV drivers in the United Kingdom, and probably as many again who drive lighter vehicles, while there are also 300,000 licensed cab and private-hire drivers. Add in 120,000 bus and coach drivers and that is at least 1.6 million people who might be put out of work by complete autonomy, even leaving out others for whom driving represents all or a substantial part of their work (people such as chauffeurs, say). Some supporters of the new technology even argue that because there is a shortage of truck drivers in some parts of the United States, autonomous technology will be necessary to keep the transport system running. Others, however, recognize that putting all those radical members of the Teamsters union out of work may create a disenchanted group who are ready to sabotage the autonomous vehicles that have replaced them through hacking – or simply by standing in front of them. This crucial aspect must be part of the debate about the future of this technology.

One of the few sober assessments of the future impact of CAVs has come from the excellent but little-noticed report from the House of Lords Science and Technology Committee cited above that was published in February 2017. Quite rightly, it highlights the fact that the government has been suckered into concentrating far too much on the ultimate utopia rather than on the far more basic realities of the introduction of autonomy:

> The Government's work on CAV for the roads sector has focused too heavily on research problems and testing technologies for highly automated vehicles with inadequate effort on thinking about deployment, especially user acceptance for road vehicles, or on the wide range of possible benefits from connected vehicles.[56]

The advent of autonomous cars is presented as a solution to many transport and other problems, but current research seems to be going down a blind alley, with its emphasis on an autonomous shared-use future that appears unreachable. The difficulties set out in the previous chapter seem insuperable, or at the very least they will require decades of development. The Lords committee recognized this and noted that the government had too readily fallen for the 'hype' around driverless cars. It recommended that research should be concentrated elsewhere:

> The Government is too focused on highly-automated private road vehicles ('driverless cars'), when the early benefits are likely to appear in other sectors, such as marine and agriculture.

It is not Luddite to question the impetus towards autonomy. However, it is essential to examine ways that the technology can be adapted to bring about societal benefit rather than profit for the auto and tech industries. The Indian government has taken the most extreme step by banning driverless cars, and while few other countries are likely to take a similarly draconian approach, many will examine more carefully the precise benefits and costs of this technology before allowing unfettered access to their roads.

All of this throws up a question about the economics of this game. Waymo is funded by Google's billions, and Microsoft, Amazon and other new tech monopolies are also investing massive sums. But even their resources are ultimately limited, and a viable business model will eventually be needed. Uber continues to be loss-making – a situation made worse by the pandemic – and it seems to be putting its long-term hopes of profit on the notion that it will eventually not need any pesky drivers, with their demands for holiday pay and sickness benefits.

So where is the money to be made? A *Financial Times* article in May 2017 set out to discover the answer:

Several engineers specialising in artificial intelligence, an area that is core to autonomous driving, told the *Financial Times* they had recently left the sector because of doubts about its viability. 'In the actual gold rush, you knew there was gold out there somewhere, and people were able to mine it,' says Josh Hartung, chief executive of Poly-Sync, which makes software for autonomous vehicles, 'In the autonomous gold rush, it's less obvious whether there is any gold there. There is effectively zero revenue that is being produced by this industry'.[57]

Hartung argued that there was a 'massive, multi-billion dollar science project that's basically on VC [venture capital] life support'. He added that even when the technology becomes ready, 'regulation, public acceptance and developing a viable business model will form barriers to widespread implementation'.

Another fallacy, therefore, is the notion that the development of this technology will come about thanks to private enterprise. The opposite is true. Not only is the American government seed funding these start-ups from a $4 billion fund set up by the Obama administration – the United Kingdom had a far more modest £100 million intelligent mobility fund created by George Osborne, which has since been expanded to around £250 million – but all sorts of regulatory provisions are being made to help their development. Governments will be intimately involved in the creation of this technology, which makes it even more important that they create the right sort of regulatory framework.

The tech companies and auto manufacturers are likely to dismiss many of the issues raised in this book.

At conferences I have attended, those involved in the development pooh-poohed my concerns and tended to emphasize how many of these problems have already been solved. 'Buses are already running driverless and cars have done millions of miles in California,' one academic insisted in a late-night conversation. But this only relates to a few tests and trials with drivers still in the vehicles. The contrast between the hype and the reality remains. We may of course reach a stage where the technical, legal, social and economic problems are all solved and a driverless future becomes a reality. But even then I suspect it will be a niche market, for tractors ploughing fields, for example, or for vehicles restricted to a set route away from the public, such as in a mine.

Even with such a limited future, we must ensure that the technology is used in the right way and enhances society rather than damages it. At the end of the day, as with so much technological progress, it is mostly the politics that will determine whether change is beneficial or not.

As this book argues, there is an enormous risk that if the development is controlled by tech companies and auto manufacturers whose sole purpose is enhancing their bottom line, then autonomous cars could be a disaster. None of this is to say we should not adopt new technology. Quite the opposite. But it needs to serve the societal interest: improving bus information systems might therefore be more useful than trying to force us into autonomous pods. There may even be a limited role for shared driverless pods, such as shuttles between airport parking and terminals: indeed, there are already pods on a guided track performing this task at Heathrow's Terminal 5.

But politicians need to be properly informed about the risks and the potential. One potential negative is if legislators and regulators feel they have to create a framework that is favourable to autonomous cars, in the same way as happened in the early days of the motor car. There is a contradiction at the heart of the happy talk from Silicon Valley about autonomous vehicles. The introduction of autonomous cars is often presented as a liberating force, providing people with greater mobility and wider access to all kinds of services. In fact, it is obvious from all the above that any move towards universal autonomy would require radical intervention from government. All kinds of restrictions would be required on other road users. Already, as we have seen, Nissan's former CEO has talked about what a pain cyclists are. The examples I have cited about people standing in front of vehicles would clearly need punitive legislation to prevent obstruction of vehicles. The Holborn problem might require huge stretches of fencing on city streets to prevent jaywalking, which is hugely damaging to the urban realm as it encourages speeding and makes walking far more difficult. That is why Boris Johnson took down miles of fencing across London when he was mayor. In other words, autonomous cars might be the catalyst for all kinds of restrictions, rather than a force for freedom. They could be the excuse to pen in pedestrians, restrict cyclists, prioritize autonomous vehicles over conventional ones, and turn cities into driverless rat-runs. Moreover, if driverless pods became the norm and made public transport uneconomic, or even killed it off entirely, what would happen to people who did not have, or did not want, access to them? To call this Orwellian seems an understatement.

In this short book I have attempted to raise numerous issues that seem to have eluded, or been ignored by, those clever people at Google, Uber, Tesla, Ford and elsewhere. Or possibly they are simply not letting on about their doubts – something that was hinted at by Rob Dingess in the previous chapter. The tale of the emperor's new clothes again seems to be the most apt analogy.

There would be enormous repercussions for auto manufacturers if the whole concept of autonomy was not realizable. While the tech companies – who are largely exploiting a monopoly position and making super-profits – can afford to lose the billions they have invested, the car companies cannot. For them, the race to autonomy is a desperate bid to survive. If, as is quite possible, the driverless car bubble bursts with as much force as the dotcom one did a decade or so ago, there will be worldwide repercussions for the industry. Tesla may well be the first to feel the impact.

I will finish with one last bit of hyped nonsense that was reported in *The Times* in 2018.[58] The paper noted that 'a study published by Highways England said that self-driving cars would reduce the need for lighting [on motorways and other major roads] because vehicles could be guided by radar'. This is, yet again, foretelling a future of *total* autonomy, which, as I have demonstrated, is probably unattainable. This shows the extent to which policy makers, who ought to be more sceptical, are beginning to tailor policies in relation to this unreachable utopia. That is the most dangerous potential consequence of all this hype. It is understandably seductive to dream of this new world, but it is also a diversion away from tackling the real and present problems posed by our transport policies, as highlighted in my previous book in this series: *Are Trams Socialist? Why Britain Has No Transport Policy*. In twenty or thirty years, perhaps, we may start thinking about turning the lights off (even though I have my doubts), but it is certainly way too early to begin considering such an action now. The fact that this is even mentioned in a report by a body such as Highways England, which is responsible for the country's trunk roads, is a result of the softening-up process that is the underlying purpose of the hype – we must not fall for it.

If the technology does become viable enough for a partial introduction, policy makers must devise the right regulatory and legal framework to ensure that other road users do not lose out. In other words, there is no reason to prioritize or give special dispensation to autonomous vehicles. Policy makers must not start implementing transport strategies on the assumption that all will be fine in a few years' time as the dream of autonomy is achieved. For example, the government has lost focus on road safety in recent years, with the result that deaths, which had declined over a long period, have started rising again: there was a 5 per cent increase in 2019 to 1,870, the highest total since 2011. It is quite conceivable that the promise of a drastic reduction as a result of autonomy could lure politicians and policy makers into neglecting continuing efforts to reduce the casualty rate – they might even be tempted into shifting resources away from implementing safety measures. What is the point of dealing with a dangerous junction if in a few years' time it will only cater for autonomous cars? Such thinking will be particularly tempting at a time of pressure on public spending. That junction could be dealt with now, with the right transport planning and regulatory policies, potentially saving lives.

Despite all the hype, despite the millions of words written about driverless cars, despite the billions of dollars being invested by the auto manufacturers and tech companies, and despite all the tests and trials across the world, nothing is really known about how this will pan out. Driverless cars could remain as purely theoretical as teleportation or personal jetpacks. The wise Lords on the Science and Technology Committee summed it up admirably:

> There is little hard evidence to substantiate the potential benefits and disadvantages of CAV because most of them are at a prototype or testing stage. Furthermore, as with any new technology or advancements, there may be unforeseen benefits or disadvantages that have not yet presented themselves.

In other words, there are simply too many known and unknown unknowns. As they concluded, 'whilst some of our evidence has suggested that CAV could have huge economic benefits, we are not convinced that the statistics provided have been properly substantiated'.

Driverless cars are pitched as an exciting technological solution to social problems. There are lots of ways in which we can reduce congestion and make the roads safer. We could simply ban all private cars from city centres, or we could charge them for using busy roads. Universal road charging would have an enormous impact on car use and would channel demand into other, more sustainable, modes of transport. Virtually all transport planners recognize that the free-for-all on roads is an insane policy, from both a planning and a social welfare point of view. Alternatively, cities could create high-frequency bus and tram networks with complete priority along main roads and then provide cheap taxi services to get people the last mile or two to their destination.

An International Transport Forum study into the patterns of travel on one day in Helsinki showed that a combination of shared-use vehicles and public transport would produce dramatic reductions in congestion and air pollution. All journeys taken on the day of the study were put through a model in which motorized road trips by private car, bus or taxi were replaced by six-seater shared taxis (which would provide an on-demand door-to-door service) and taxi-buses (which would offer a street-corner-to-street-corner service booked half an hour in advance). The effects were far-reaching, with a fall by a third in both CO_2 emissions and congestion. All of the car journeys currently undertaken in the Helsinki metropolitan area could have been provided with just 4 per cent of current private vehicles. As the report authors concluded:

> Much of public parking space could be used for other purposes. Shared mobility also means fewer transfers, less waiting and shorter travel times compared to traditional public transport. The improved quality of the service could attract car users that currently do not use public transport and foster a shift away from individual car travel.[59]

In other words, the same sort of gains that are being promised by the advocates of autonomous cars could instead be obtained by restrictions on private-car use and the provision of new types of service. This completely undermines the arguments in favour of autonomy. It is not the technology that can deliver the changes they forecast, it is transport policy, irrespective of any shift to autonomy. The autonomous-vehicle lobby is pushing for a revolutionary shift in car ownership patterns, but this could equally be achieved now, and the reason it has not been is nothing to do with technology and everything to do with politics. People want their own cars since they are allowed to drive them everywhere freely. Take that right away, such as through banning vehicles in major city centres, and they may well all sign up to Uber or its rivals. But they could do that now, without needing the vehicles to be driverless. The push for autonomy is borne of a desire to promote a technology that will be highly profitable for the tech and auto companies – it has nothing to do with freeing up road space or making life better for road users.

The examples cited above show that there are alternatives. It is only because transport policy has been geared towards catering for individual car use that our cities are so congested and polluted. It is unclear how driverless cars will help address any of the fundamental problems created by our car culture. Even their most fervent supporters admit they may not necessarily reduce congestion because of the zombie kilometres they will drive with no one inside them. This short book has only touched on the myriad obstacles needed to create the driverless future envisaged by its proponents. The glib assumption that autonomous cars are inevitable must be challenged.

Politicians and policy makers should not be at the mercy of technology. They should be the masters of it, harnessing it to their political aims. That must be the starting point when addressing the potential advent of autonomous vehicles. They should also be prepared for it not happening, and they should apply themselves to other, more realistic, ways of addressing the problems cars cause us.

Postscript

Since publishing the first edition of this book two years ago, a lot has happened in the world of autonomous vehicles. Numerous new trials have taken place, while other promised tests have never materialized; a (sort of) robotaxi service has been launched by Waymo in Phoenix, Arizona; and Elon Musk, the boss of Tesla, made a promise that a million cars would be available for driverless robotaxi use (though this was clearly a fantasy when several Teslas had, in fact, been involved in accidents related to overdependence on their 'autopilot' system). The hype has abated somewhat in the face of the realities of the technology and the lack of enthusiasm among the general public, but many optimistic predictions continue to be made. Perhaps most astonishingly, in the spring of 2020 Waymo raised $3 billion in outside investment for its continued development of the concept.

Despite all this, it is inarguable that the central premise of this book has been reinforced by the events of the past two years. There is at present no sign of any commercial use of autonomous vehicles, there is no clear business model, and there are increasing doubts about whether the very concept is even feasible. The technology has reached Level 2 in production vehicles, according to the Insurance Institution for Highway Safety (IIHS),[60] which essentially means that there is some partial automation of minor tasks such as lane changing, but drivers have to remain alert at all times to prevent accidents. Some trials claim to be at Level 4 but these are very limited: mostly they still have safety drivers and only operate in good conditions.

The case of Tesla is particularly concerning. Tesla's 'autopilot' is no such thing, and overreliance on it has been the cause of numerous accidents. Indeed, Elon Musk's insistence that his vehicles are 'self-driving' has been widely criticized by safety authorities in several countries, and unsurprisingly there have been numerous lawsuits in which complainants allege that the company's misleading claims on self-driving have led to fatalities and injuries. A video that went viral in June 2020 showed a Tesla smashing into an overturned truck on a Taiwanese highway, despite having plenty of time to take avoiding action, because it was on 'autopilot'.

There have also been several reports of people assuming that the 'autopilot' facility can be used to drive them home safely when they are too drunk to take the wheel themselves. In July 2020, the driver of a Tesla whose 'Autopilot' was turned on when it smashed into the back of a stationary ambulance was arrested under suspicion of 'driving under the influence'.

There are numerous other examples that reveal that the technology may not be the boon to safety that is claimed by its promoters. As we have seen, supporters of the concept have claimed that since 94 per cent of accidents are caused by human error, eliminating the driver would cut their number drastically, or even entirely. Ignoring the fact that it is highly likely that software errors will add a different type of risk, the very basis of this claim has been challenged by research from the IIHS,[61] which has calculated that just a third of accidents would be prevented by the use of autonomous vehicles. This is because only accidents that are what the researchers call 'sensing and perception' errors, such as driver distraction or failure to spot a hazard, will be prevented. The technology cannot prevent the majority of accidents, which the IIHS believes are caused by 'prediction errors', such as misjudging the speed of other vehicles, excessive speed when road conditions are treacherous, and mistaken driver efforts to avoid a crash. One example is when a cyclist swerves into the path of an autonomous car. The vehicle may have seen the cyclist but it can't manoeuvre quickly enough to avoid hitting them.

Cyclists are in fact a particularly difficult hazard for autonomous vehicles to perceive properly, and it was no coincidence that the first fatality caused by a 'driverless' car was a cyclist – or, more specifically, a woman pushing a bicycle on a highway. The death of Elaine Herzberg in March 2018 has become a defining moment for the autonomous vehicle industry and has severely dented the public's confidence in the concept. Herzberg was pushing a bicycle laden with shopping bags across a high-way and was hit by a prototype Uber self-driving Volvo that had a safety driver but was operating in autono-mous mode. The car's software was clearly confused by the particular combination of human, bicycle and shop-ping bags. The car's system had detected Herzberg six seconds[62] before the crash but had initially classified her as an unknown object, then as a vehicle, and finally as a bicycle, each of which required a different response. By then it was too late to alert the safety driver – who was, in any case, not paying sufficient attention – to make an emergency stop. Herzberg was killed by the impact.

While the failure of the 'driver' to intervene was a contributory factor, the accident highlighted the fact that the technology is nowhere near as safe as has been claimed and that making it safe than conventional vehicles is a far bigger task than previously realized. As a result of the accident, Uber suspended all its autono-mous vehicle testing for the rest of the year and has sub-sequently only reinstated a more limited programme. And while the 'limited' programme is still costing the company $2 billion per year, Uber's management has been less publicly optimistic about the potential to develop a fully driverless vehicle in the near future.

While the Herzberg accident resulted in a great deal of reappraisal of their testing programmes by many of the companies involved in autonomous vehicle research, wild claims about the technology are still regularly being made. For example, Zenzic, the organization that coordinates the United Kingdom's research programme, claimed in a press release[63] that connected autonomous vehicles could reduce transport emissions by between 5 per cent and 20 per cent by reducing congestion and that the technology was 'the key to becoming climate neutral' – a tenuous argument given that the widespread use of autonomous vehicles might lead to many more cars on the road.

The environmental claims are based on the 'triple revolution' of shared use, driverless and electric, which is the model presented by supporters of autonomous vehicles. This concept was the centrepiece of much of the hype and, as I have argued, it was necessary because without shared-use vehicles, there would be an increase, rather than a decrease, in traffic in a world dominated by driverless cars. Autonomous vehicles would in many ways encourage greater car use, e.g. by enabling blind people 'to drive', and there is also the prospect of large numbers of empty vehicles driving around, leading to a significant increase in congestion.

Even pre-pandemic, the technical difficulties that needed to be overcome before this completely new transport scenario could be reached had already led to a reduced emphasis on this model, and, in particular, the shared-use aspect of the concept has now virtually been killed off altogether because of the easy transmission of

Covid-19 in confined spaces. Who would want to share the small enclosed space of a taxi with a stranger and risk catching a deadly disease? Even if a vaccine and treatment for Covid-19 are developed, fears of similar illnesses are likely to reduce the potential of shared-use vehicles for some time to come.

Zenzic boldly claims that the industry will be worth £52 billion in the United Kingdom and £907 billion worldwide by 2035. This figure is taken from a 2017 Market Forecast report by the Connected Places Catapult,[64] a government research organization supported by Zenzic. Although Zenzic and other supporters of the technology use the figures widely, they are strongly caveated in the original Catapult Report:

> Despite the significant surge in interest in this sector in recent years, CAVs and CAV technologies are yet to be fully developed, and an industry consensus around factors such as costs and consumer attitudes has yet to emerge. Therefore, the accuracy of the forecasts set out in this study are inevitably limited by uncertainties around adoption rates, costs and labour intensities for these technologies.

In other words, these predictions are not really worth the paper they are (not) printed on. My favourite piece of analysis is from a 2018 report by BDO, an accountancy and business advisory company. It summarized the predictions of eighteen car makers and suggested – with a precision that presumably reflects an element of sarcasm – that 'driverless technology will be ready around 2am on June 11th, 2021', whereas 'ride-hailing services and technology suppliers predict that driverless car technology will be ready by midnight, March 14th, 2020'.[65]

The extent and breadth of the claims for autonomous vehicles have led to the development of the concept of 'autonowashing', which is akin to the 'greenwashing' used to describe the fake environmental commitments often made by corporate entities. In an article for the online academic magazine *Transportation Research Interdisciplinary Perspectives*,[66] Liza Dixon argues that much of the material put out by the companies that are developing autonomous vehicles is misleading as it fails to distinguish properly between autonomy and driving aids.

She defines autonowashing as making unverified or misleading claims that misrepresent the appropriate level of human supervision required by a partially or semi-autonomous product, service or technology. She cites the use of vague language and the failure to prove claims as being characteristics of autonowashing, and she highlights the media's guilt when it comes to its 'utopian' reporting and exaggeration of the level of autonomy. Indeed, there are numerous examples of articles whose headlines suggest they are about 'driverless' vehicles but that go on to reveal that there is a safety driver at the wheel and sometimes an engineer on the back seat, too.

Autonowashing is a useful concept and term, and one that is worthy of greater use. And as Dixon points out, the phenomenon is somewhat self-defeating for the industry, which depends on building trust among potential users. By exaggerating claims and failing to consider disadvantages, the industry is weakening its own case. She writes:

> Autonowashing leads to overtrust, which leads to misuse. If a driver management system is unable to assist the user in error prevention, accident, injury or death may occur. This results in negative media coverage which can then stir public distrust in vehicle automation, threatening the return on investment.

She is particularly critical of Tesla for portraying Level 2 automation as 'self-driving' and for using the misleading term 'autopilot'. Part of the problem, she posits, is the established six-level grading system for autonomy. It may in fact be best to simply have a binary system: driverless or not driverless.

Most serious players in the industry are now aiming for Level 4. At this level, vehicles can drive themselves in certain settings without a driver but it is still possible for a human controller to take over. This is in contrast with Level 5, which represents true full autonomomy: go anywhere, anytime. Stopping at Level 4 tacitly rules out the nirvana of a world, or city, dominated by driverless pods that are able to function in all circumstances and all weather conditions.

The hype around the issue has, in one respect at least, borne fruit for the auto manufacturers and tech companies who are so eagerly and expensively still promoting the concept of autonomous vehicles. Supported by fierce industry lobbying, American legislators are expected to pass a series of bills in Congress that would allow manufacturers and tech firms to deploy self-driving vehicles on roads without having to adhere to existing safety standards. The legislation being considered would also prevent states from passing their own laws regulating driverless cars, which the industry has argued is necessary to avoid a patchwork of different rules across the United States. London, too, is expected to see the first tests of 'driverless' cars on its streets in 2021, although there are as yet no specific regulations. These are being developed by the Law Commission, which is due to publish the results of its final consultation exercise in 2021, with the idea that this will form the basis of future legislation.

Over the past couple of years, some industry predictions have become more cautious, and there is often now greater awareness, and even sometimes transparency, about the obstacles and difficulties that need to be circumvented. Several carmakers and technology companies have concluded that producing truly autonomous vehicles is going to be harder, slower and more costly than they previously thought.

For example, as reported by the *New York Times* on 17 July 2019, Ford's chief executive Jim Hackett said in a speech in Detroit: 'We overestimated the arrival of autonomous vehicles.' The same article went on to quote Bryan Salesky – the chief executive of a start-up autonomous vehicle technology research company called Argo – who said that the arrival of these vehicles was 'way in the future'. He admitted that while many companies like his own had developed about 80 per cent of the technology needed to put self-driving cars into routine use, 'the remaining 20 per cent, including developing software that can reliably anticipate what other drivers, pedestrians and cyclists are going to do, will be much more difficult'.

This kind of handwringing is becoming widespread, as doubts have crept in and certainties have become, well, less certain. In the early days of the development of autonomous vehicle technology, there was an assumption that 'artificial intelligence' would solve all the difficult issues around safely identifying objects and situations. The limitations of AI are becoming apparent, though, and are being exposed by its alternative name: machine learning.

One issue that has emerged is the sheer difficulty of teaching the machines about *every* possible eventuality: the problem of the lion walking down Main Street, say. Not a common occurrence, perhaps, but one that will certainly confuse the computer. It seems there will always be outliers that have not been considered, as poor Ms Herzberg discovered to her cost. Eliminating all of these is an almost insuperable problem.

Indeed, some sceptics of the technology argue that it can only be achieved through extensive testing. Michael DeKort, a former aerospace engineer turned whistle-blower, argues that this makes it impossible to create a fully autonomous vehicle:

> It is not possible, neither in time or money, to drive and redrive, stumble and restumble on all of the scenarios necessary to complete the effort. The other problem is that the process will cause thousands of accidents, injuries and casualties when efforts to train and test the AI move from benign scenarios to complex and dangerous scenarios.[67]

In other words, many of the test vehicles that are needed will be involved in accidents caused by AI failures, and, he goes on to argue, this will become socially unacceptable, leading to the abandonment of the whole driverless project. He recommends, instead, testing on simulators, but this too would have to be very extensive and expensive.

An example of the limitations of the technology that backs up DeKort's thesis is the revelation – in research undertaken at Georgia Tech[68] – that the computer programs that are in use tend to find it harder to identify dark-skinned people. The reason for this is that the AI has been developed on the basis of street scenes and other public scenarios that have relatively few black people in the frame, meaning that the machines are not as good as identifying them as they are at identifying white people.

Even more significantly, as well as greater scepticism in media coverage of the progress of autonomous vehicle development, there seems to be less confidence about technological progress within the industry itself. In an interview with the author, Liza Dixon suggested that in the past couple of years, 'progress seems to have stalled as there has been no sign of a game changing breakthrough'. An article on CNET[69] in November 2018 quoted the CEO of Waymo, John Krafcik, as expressing doubts over whether autonomous cars would ever become ubiquitous:

> It'll be decades before autonomous cars are widespread on the roads – and even then, they won't be able to drive themselves in certain conditions. Autonomy always will have some constraints.

While this suggests that there is a need for a model that is very different from the ones previously proposed, there is no sign at this stage of what it is.

The problem seems to be that there has been so much progress in so many aspects of technology that a belief that anything is achievable has grown. However, the sheer complexity of developing true autonomy is only just beginning to be understood, even by those, like Waymo, who have worked in the field for a decade.

Despite Waymo's success in obtaining outside investment, there are signs that the driverless car boom is on the wane, at least in terms of the valuation of companies in the industry. At its height, in 2016, General Motors acquired Cruise – a three-year-old, forty-person start-up – for $1 billion (including performance incentives). A few months later, Uber paid $680 million for Otto, a six-month-old autonomous trucking start-up. That worked out at $10 million per engineer: a level that is no longer attainable.

The sums going into the industry remain staggering, nevertheless, especially given that there is no end product on the horizon. A survey of the top thirty companies in the field published in *The Information* revealed that $16 billion was spent on autonomous vehicle R&D in 2019:

> Just three companies spent half of that money – Alphabet's Waymo, GM's Cruise and Uber... Four other companies, including Apple, Baidu, Ford and Toyota, spent most of the rest.[70]

In the United Kingdom, by the middle of 2020 the government had promised £200 million to private companies to develop autonomous vehicles under the auspices of Zenzic, and there was the expectation that further money would be made available. A feature of this investment programme is that very little of it is publicized, and apart from a few glossy presentations and PR launches, few assessments of progress are made public.

Zenzic has devoted much energy to producing its 'UK Connected and Automated Mobility Road Map to 2030': a 106-page document of Byzantine complexity that purports to show the necessary steps to develop full connectivity and autonomy by 2030. The report

> provides direction for decision makers, investors and policy makers for the mobile future. The roadmap is a tool, created by and intended for multiple sectors, forging new relationships and achieving collaboration across industries. With a single vision of interdependencies, the roadmap addresses developments needed to achieve connected and automated mobility at scale by 2030.[71]

However, in order to reach this Holy Grail, dozens of milestones – ranging from legal and regulatory issues to technical and social factors – need to be met, and they are all interdependent and subject to a wide variety of assumptions.

Adding to the sea of uncertainty regarding autonomy is the Covid-19 pandemic and its aftermath. As I mentioned earlier, shared use is likely to be less desirable, at least in the short term, but crucially so is public transport. More people will undoubtedly have found that working from home, perhaps for part of each week, is not only feasible but desirable. Employers may become more trusting of allowing staff to do so. The medium-term, let alone the long-term, implications of this trend are unclear, but I think they suggest that public transport may become less economic and that the roads may be slightly easier to negotiate at commuting times, although not if public transport users shift to their cars.

The promoters of driverless car have argued that the reaction to the pandemic may prove to be beneficial, as people are likely to prefer a robo-taxi to ones with a driver, but there are many counterarguments to this, notably that there would be no one to clean and disinfect the vehicle between rides. The focus on Covid-19 has, frankly, rather highlighted the irrelevance of technology to solving society's major problems, and there may well be less interest in such futuristic concepts in the coming years when humanity has shown that it cannot even deal with a simple virus effectively.

Despite all the doubts, the accidents, the pandemic and some belated recognition of the obstacles to full autonomy, just as I was writing these final sentences Amazon announced that it was spending $1.2 billion to buy Zoox, a self-driving start-up founded in 2014 that aims to develop a fully fledged robo-taxi.[72] That is a trivial amount for Jeff Bezos, of course, but it demonstrates the resilience of the driverless dream. And this is the point at which it becomes almost impossible to explain this story in rational terms. Even after closely following progress in the industry for many years and having had conversations with literally hundreds of people, no one has been able to explain to me how this readiness to invest yet more billions into developing this technology follows any sort of logic within the framework of conventional capitalism. Even recognizing that the likes of Google, Apple and Amazon are making super profits – in other words, their surpluses are the product of monopoly positions in the market rather than the fruits of any conventional profit to investment ratio – and therefore have money to burn, there are many other investors who are also throwing vast sums at a vague concept that has no business case.

According to a *Fortune* magazine article of 7 January 2020, while Waymo is by far the market leader after eleven years of research, the company 'remains an expensive science project in search of a business'. It is quite extraordinary that after billions of dollars of spending, no realistic business model has been put forward by any of the key players. There have been glossy illustrations of potential uses, but no worked-out model of how money can be made from this technology has been advanced. The focus of both Amazon and Waymo has shifted to the carriage of cargo, but while it may be possible to imagine that long-distance trucks running on highways could become driverless and may even platoon – though there are still great technological barriers to overcome – is this really likely to save enough money to make the massive investment in the technology worthwhile? Moreover, surely Amazon, despite its ludicrous attempts to develop drone technology, must realize that residential deliveries will always require a human to knock on the door and take the package out of the van. And if there has to be a person on the van, they may as well drive it. Again, there is no logic to be found here.

And nor is there any logic to Uber's now rather slower march towards robo-taxis. Under the current model, the driver owns the vehicle, so Uber would need to raise vast amounts of capital to pay for its driverless taxis, all to save the cost of a low-paid driver.

Moreover, there is now even less trust in this technology than there was when the first edition of this book was written. Nearly half of Americans say they would not get in a self-driving taxi, according to a poll by the advocacy group Partners for Automated Vehicle Education.[73] The poll, carried out in early 2020, found that 48 per cent of the 1,200 adults surveyed would 'never get in a taxi or ride-share vehicle that was being driven autonomously', while a further 21 per cent said they were unsure about doing so. While a fifth of respondents said that autonomous vehicles would never be safe, another fifth stated, incorrectly, that it is possible 'to own a completely driverless vehicle today', highlighting the confusion that still remains over how far the technology has already developed. On the other side of the coin, people want to continue driving. A post-lockdown survey in *Le Monde* found that half of all car owners actually missed driving while they were unable to travel.

The Covid-19 pandemic has added further woes to the automotive industry. Already struggling with overcapacity and battling against increasing hostility due to greater awareness of the environmental costs of their products, the industry now faces the prospect of a global recession. This may, though, create the opportunity for a rethink. The billions poured into autonomous vehicle R&D has partly been a defense mechanism: what if the other guys get there first? What if this represents the industry's only future? The investment was therefore seen as a necessary part of their survival strategy. This could now change. With even less money available for research, it would be sensible for car manufacturers to devote their funds to developing electric vehicles, which are undoubtedly a growth area, rather than autonomous ones, for which, as we have seen, there is no business model and a very uncertain future.

And yet few of these considerations are properly assessed in the very expensive analyses produced by consultancy firms and City traders. There is, in fact, a quasi-religious fervour to this debate. None of the leading protagonists in the United Kingdom will speak to me any more because I have questioned their faith. I have participated in numerous debates at conferences on the future of driverless cars and my opponents invariably come across with all the devotion of one of those besuited evangelists who try to stop me outside Tube stations to ask me to pray for Jesus. There may be dozens of pins already in the bubble, but somehow it still refuses to burst.

Endnotes

1. S. Parissien. 2013. *The Life of the Automobile: A New History of the Motor Car*, p. 186. Atlantic Books.

2. Quoted in L. Sloman. 2006. *Car Sick: Solutions to Our Car-Addicted Culture*, p. 12. Green Books.

3. Sloman (2006, p. 11).

4. C. Wolmar. 2016. *Are Trams Socialist? Why Britain Has No Transport Policy*, p. 38. London Publishing Partnership.

5. G. Lean. 1999. Change at the pumps: the slow death of lead. *Independent on Sunday*, 26 December (https://ind.pn/2AVU56R).

6. B. Ji. 2010. GM: glimpse into auto future. *China Daily*, 1 May (http://bit.ly/2hAz4q1).

7. L. Tillemann. 2015. *The Great Race: The Global Quest for the Car of the Future*, p. 17. Simon & Schuster.

8. Tillemann (2015, p. 18).

9. All these quotes come from a statement given to the Committee on Commerce, Science, and Transportation of the United States Senate on 15 March 2016 (http://bit.ly/2zHXoko).

10. See note 9.

11. Agence France-Presse. 2016. Uber launches driverless car service. *Daily Telegraph*, 14 September (http://bit.ly/2cVA7kV).

12. See note 11.

13. C. Kang. 2017. Pittsburgh welcomed Uber's driverless car experiment. Not anymore. *New York Times*, 21 May (http://nyti.ms/2rai2Fl).

14. S. Curtis. 2017. Domino's launches ROBOT pizza deliveries in Europe. *Mirror*, 30 March (http://bit.ly/2nnMjLP).

15. O. Garret. 2017. 10 million self-driving cars will hit the road by 2020. *Forbes*, 3 March (http://bit.ly/2hGUG7s).

16. S. Hill. 2017. What Dara Khosrowshahi must do to save Uber. *New York Times*, 30 August (http://nyti.ms/2wMsAoj).

17. Driverless Car Market Watch website (http://bit.ly/1eQzhBL).

18. H. Sanderson. 2017. Electric car demand sparks lithium supply fears. *Financial Times*, 9 June (http://on.ft.com/2r8dKPM).

19. A diagram explaining the levels can be found at http://bit.ly/2dm3mYI.

20. Reuters. 2016. Elon Musk says Tesla's new autopilot likely would have prevented death. *Fortune*, 12 September (http://for.tn/2c3lV3n).

21. M. Posky. 2017. Autonomous features are making everyone a worse driver. The Truth About Cars website, 10 August (http://bit.ly/2jxwncJ).

22. T. Harford. 2016. *Messy: The power of Disorder to Transform Our Lives*, p. 183. Riverhead Books.

23. A. Fraher. 2017. US Navy collisions point to the risks of automation on sea, air and land. The Conversation website, 30 August (http://bit.ly/2hGWAVK).

24. Harford (2016, p. 186).

25. Interview with the author.

26. House of Lords, Science and Technology Select Committee (2nd Report of Session 2016–17). 2017. Connected and autonomous vehicles: the future? House of Lords Paper 115 (http://bit.ly/2f1aBbK).

27. K. Naughton. 2017. Ford's dozing engineers side with google in full autonomy push. Bloomberg Technology website, 17 February (https://bloom.bg/2kFiYKQ).

28. See note 9.

29. R. Vierecki *et al.* 2016. Connected car report 2016: opportunities, risk, and turmoil on the road to autonomous vehicles. Strategy&, 28 September (https://pwc.to/2wUs3IT).

30. Interview with the author.

31. K. Piper. 2020. It's 2020. Where are our self-driving cars? *Vox*, 28 February (https://bit.ly/38Ifqlv).

32. Forecasts. Driverless Car Market Watch website (http://bit.ly/1eQzhBL).

33. D. Fagella. 2017. Self-driving car timeline for 11 top automakers. VentureBeat website, 4 June (http://bit.ly/2rDv4ZK).

34. D. DeGasperi. 2017. Hyundai planning driverless car by 2020. Drive website, 20 February (http://bit.ly/2AWsHWj).

35. E. Behrmann. 2016. Volvo cars plans a self-driving auto by 2021. Bloomberg Technology website, 22 July (https://bloom.bg/2hzUEuJ).

36. Quoted in Dingess (2017).

37. See note 33.

38. See note 37.

39. Dingess (2017).

40. See note 9.

41. D. Shepardson. 2017. Tesla driver in fatal 'Autopilot' crash got numerous warnings: US government. Reuters, 19 June (http://reut.rs/2zbiHeF).

42. S. Garfinkel. 2017. Hackers are the real obstacle for self-driving vehicles. *MIT Technology Review*, 22 August (http://bit.ly/2vU2Cqe).

43. A. Greenberg. 2017. Securing driverless cars from hackers is hard. Ask the ex-Uber guy who protects them. *Wired*, 12 April (http://bit.ly/2p5bUwY).

44. Dingess (2017).

45. Top misconceptions of autonomous cars and self-driving vehicles. Driverless Car Market Watch website (http://bit.ly/22r3FqO).

46. Quoted in C. Reid. 2017. Nissan driverless car guilty of 'close pass' overtake of UK cyclist. BikeBiz website (http://bit.ly/2maBbno).

47. G. Topham. 2017. Semi-automated truck convoys get green light for UK trials. *The Guardian*, 25 August (http://bit.ly/2gamKPw).

48. A. Mamiit. 2014. Rain or snow? Rock or plastic bag? Google driverless car can't tell. *Tech Times*, 2 September (http://bit.ly/2za1HWa).

49. D. Wakabayashi. 2017. Where driverless cars brake for golf carts. *New York Times*, 4 October (http://nyti.ms/2yY1y4s).

50. Interview with the author.

51. Smart Transport. 2020. Half of drivers do not trust driverless technology, finds InsuretheGap.com. Smart Transport website, 21 April (https://bit.ly/3222sO2).

52. See note 42.

53. M. Bridge. 2017. How driverless cars will make our lives better. *The Times*, 8 April (https://bit.ly/2054FRq).

54. All these quotes are from an interview with the author.

55. See note 26.

56. See note 26.

57. L. Hook and T. Bradshaw. 2017. Driverless cars inspire a new gold rush in California. *Financial Times*, 23 May (http://on.ft.com/2qiE2cB).

58. G. Paton. 2017. Self-driving cars could run on unlit roads to conserve energy. *The Times*, 9 October (http://bit.ly/2zMycZo).

59. International Transport Forum. 2017. *Shared Mobility: Simulations for Helsinki*. OECD (http://bit.ly/2zjWxTa).

60. IIHS. 2019. New studies highlight driver confusion about automated systems. Insurance Institute for Highway Safety website, 20 June (https://bit.ly/2CoWfks).

61. Road Traffic Technology. 2020. Self-driving cars may prevent only one-third of crashes, says study. Road Traffic Technology website, 5 June (https://bit.ly/2CrDigY).

62. Preliminary report by the National Transportation Safety Board, 24 May 2018.

63. J. Muir. 2020. Connected vehicles key to achieving net zero? Autonomous Vehicle International website, 21 May (https://bit.ly/2AK2Zce).

64. Transport Systems Catapult. 2017. Market forecast: for connected and autonomous vehicles. Transport Systems Catapult report, July (https://bit.ly/3iNXr1C).

65. S. Francis. 2018. Automakers' own predictions on when driverless cars will arrive. *Robotics and Automation News*, 25 July (https://bit.ly/2W6WxDB).

66. L. Dixon. 2020. Autonowashing: the greenwashing of vehicle automation. *Transportation Research Interdisciplinary Perspectives*, Volume 5 (May), article 100113 (https://bit.ly/3gDSExY).

67. M. DeKort. 2018. Medium.com article, 8 March (https://bit.ly/3futcuN)

68. K. Wilson. 2020. Study: AVs may not detect darker-skinned pedestrians as often as lighter ones. StreetsBlog USA, 17 June (https://bit.ly/2ZiY4su).

69. M. Gurman. 2018. Waymo CEO says self-driving cars won't be ubiquitious for decades. Bloomberg Technology, 13 November (https://bloom.bg/3eiNiGT).

70. A. Efrati. 2020. Money pit: self-driving cars' $16 billion cash burn. *The Information*, 5 February (https://bit.ly/2ALDlnw).

71. Zenzic. 2019. UK connected and automated mobility roadmap. Zenzic UK report (https://bit.ly/2WoN0Oz).

72. K. Hale. 2020. Amazon speeds towards $1.2 billion self driving black-led car company Zoox. *Forbes*, 7 July (https://bit.ly/2W5CMfS).

73. I. Boudway. 2020. Americans still don't trust self-driving cars, poll shows. *The Star (Technology)*, 19 May (https://bit.ly/2ZeKIo8).

Photo credits

'A Google self-driving car at the intersection of Junction Ave and North Rengstorff Ave in Mountain View' (page 19). By Grendelkhan (own work) [CC BY-SA 4.0 (http://creativecommons.org/licenses/by-sa/4.0/)], via Wikimedia Commons.

'Uber's self-driving car test driving in downtown San Francisco' (page 24). By Diablanco (own work) [CC BY-SA 3.0 (https://creativecommons.org/licenses/by-sa/3.0/deed.en)], via Wikimedia Commons.

'Levels of automation' (page 41). Created for this book by the publisher.

'In 100 metres turn left (but only if you are on foot)' (page 51). By David Stowell (from geograph.org.uk) [CC BY-SA 2.0 (https://creativecommons.org/licenses/by-sa/2.0/deed.en)], via Wikimedia Commons.

'431ch twin turbo 3.0 550N.m 1572kg 0-100 en 4.1sec à partir de 82 300 €' (page 64). By Falcon® Photography from France [CC BY-SA 2.0 (https://creativecommons.org/licenses/by-sa/2.0/deed.en)], via Wikimedia Commons.

'Looking northeast as bicyclers go down 5th Avenue after a snowfall' (page 83). By Jim Henderson (own work) [CC0 1.0 Universal (https://creativecommons.org/publicdomain/zero/1.0/deed.en)], via Wikimedia Commons.

'A massive pothole in Carisbrooke Road, Newport, Isle of Wight, so big it had to be sectioned off to prevent cars getting damaged by driving over it' (page 89). By Editor5807 (own work) [GNU Free Documentation License, Version 1.2 (https://commons.wikimedia.org/wiki/Commons:GNU_Free_Documentation_License,_version_1.2)] via Wikimedia Commons.